Teaching STEM to First Generation College Students

Teaching STEM to First Generation College Students

A Guidebook for Faculty & Future Faculty

Gail Horowitz

INFORMATION AGE PUBLISHING, INC.
Charlotte, NC • www.infoagepub.com

Library of Congress Cataloging-in-Publication Data

A CIP record for this book is available from the Library of Congress
http://www.loc.gov

ISBN: 978-1-64113-596-2 (Paperback)
      978-1-64113-597-9 (Hardcover)
      978-1-64113-598-6 (ebook)

Printed in the United States of America

Contents

Preface

How to Use This Book

This book is intended as a practical guide for STEM faculty and future faculty like you, who work with first-generation college students. This book will provide you with easy to implement strategies that you can utilize in your teaching, advising, and mentoring of first-generation students.

The primary focus of this book is on how you can best help first-generation STEM students be successful in the classroom. But this book does not ask you to revamp your courses or revolutionize your teaching styles and pedagogies. Rather, it explains how you can help students learn to adopt more effective study strategies by embedding appropriate information about study strategies into your classrooms.

Chapter 1 introduces you to two very different students, one, a first-generation student and one, a continuing generation student. Chapter 2 clarifies the multiple ways in which the term first-generation college student is used. Chapter 2 describes the demographics of first-generation college students and discusses what is understood about their trajectories and outcomes in higher education.

Chapter 3 introduces the main argument of the book, that the key to success for first-generation STEM students is the development of sophisticated study skills; namely, well-developed *self-regulated learning* abilities. Chapter 4

Teaching STEM to First Generation College Students, pages ix–x
Copyright © 2019 by Information Age Publishing

argues that training students to improve their *self-regulated learning* abilities can easily be done without sacrificing class time or curricular content.

For those who are looking to jump ahead to the *most practical, nuts and bolts* part of the book, Chapter 5 is a hands-on guide as to how you can incorporate ideas of self-regulated learning (improving study skills) into all aspects of your courses (e.g., syllabi, lectures, office hours, etc.). Similarly, Chapter 6 is a *practical* guide for *graduate student instructors,* particularly international graduate students, addressing how they can work most effectively with first-generation STEM students.

For those who want to *dig deeper,* Chapter 7 examines a range of psychological factors that impact the learning behaviors of first-generation students, while Chapter 8 discusses how you can effectively mentor first-generation students, especially those who are dealing with academic and/ or personal challenges.

The last chapter of the book, Chapter 9, is for *undergraduates.* It is intended as a chapter of advice to be shared with first-generation STEM students who find themselves struggling in their STEM courses, but are unsure why they are struggling and experience self-doubt. Chapter 9 helps students better understand what they are going through and what is happening to them, but it is also provides *practical* and direct guidance to them as to how they can become more effective self-regulated learners.

Acknowledgments

This book would never exist were it not for my students, particularly the students of CUNY-Brooklyn College. It is through listening to their stories that I have learned, day by day, to become a better teacher and better mentor of first-generation STEM students.

My teaching has also been stimulated, influenced, and impacted by the numerous colleagues I have had the pleasure of knowing over the years, colleagues across disciplines at Yeshiva University and CUNY-Brooklyn, as well as researchers in the fields of science education, chemical education, motivation, and self-regulated learning. I am grateful to my students and my colleagues who have made this book possible and who have enriched and enhanced its development.

1

Introduction

Different Students, Different Stories

It is late morning on a cold Monday midway through the spring semester. A student drops by my office to talk about how she is doing in my organic chemistry lecture course. The student looks vaguely familiar, but I cannot recall her name, and I don't think she has ever been to my office hours before. I notice that the student is soft spoken and conservatively dressed, and that she asks permission before taking a seat in one of the chairs facing me. When I ask how she did on the midterm, the student admits that she failed it miserably. She explains: She came into the exam feeling like she really understood the material, but somehow all her knowledge evaporated when the test began. I ask how she allocates her study time. She replies that she spends most of her time reading the textbook, but that sometimes if she is struggling to understand a topic, she supplements her reading by watching a video on Khan Academy. We talk further and I make some suggestions about how she can modify her study habits so as to be more successful.

A few hours later, another student drops by. This is a student I know well, and I greet her by name. This student is stylishly dressed and enters with an

Teaching STEM to First Generation College Students, pages 1–5
Copyright © 2019 by Information Age Publishing

air of confidence. Today, as usual, she arrives with her notebook in hand, full of the homework problems she is working on. The student quickly flips to the problems with which she is having difficulty, readily finding them because she has marked them with yellow highlighter. One by one, the student goes through the problems she has questions about. In one case, I am able to clear up a misconception she is having. In another case, I clarify that a particular question she is stuck on is something we haven't yet covered. (I am running a bit behind this semester.) Before she leaves, the student mentions that she feels like she needs extra practice on one topic from Chapter 6 and I quickly direct her to some extra practice problems online.

Back in 2010 when I started teaching at Brooklyn College (a public, urban university located in New York City), if you had asked me why these students differed so much in their study habits, I probably would have just given you a puzzled frown. But by now, I am not at all surprised when the first student tells me that both her parents are blue collar workers and that neither has had any college education. Nor am I surprised to learn that the second student has a parent with a graduate level education and a professional occupation.

Whose Fault Is It?

At institutions like mine, failure rates in introductory STEM courses are often as high as 50% (and sometimes even higher). Not surprisingly, tensions run high among various stakeholders at these institutions, with each looking for someone other than themselves to blame for these depressing statistics. Some blame the faculty (e.g., for poor or antiquated teaching methods). Others blame the institutions (e.g., for large class sizes, a lack of resources, an overemphasis on research, and/or an overreliance on adjuncts and teaching assistants). But often it is students who are subject to much of the blame: for having unrealistic expectations about college, for being poorly prepared from high school, for being lazy or amotivated and/ or for being too caught up in electronic gadgets and social media.

Retention of Underrepresented Groups in STEM

Much research has been devoted towards understanding why some students drop out of the STEM pipeline (Gasiewski, Eagan, Garcia, Hurtado, & Chang, 2012; Hunter, 2016; Malcom & Feder, 2016; Seymour & Hewitt, 1997) and towards ascertaining which teaching methods can most effectively improve the retention and performance of these students (Eberlein et al., 2008; Freeman et al., 2014; Kober, 2015). Additionally, although much has been written about the risk factors, trajectories, and outcomes of women and ethnic minorities in STEM (Chang, Eagan, Lin, & Hurtado, 2011; Hurtado, Newman, Tran, & Chang, 2010; Russell & Atwater, 2005; Treisman, 1992), researchers are now beginning to explore the experiences of "first-generation" college students in STEM (students who are the first-generation in their family to attend or complete college; Davis, 2010).

First-Generation College Students in STEM

Examining the experiences of first-generation students and what can be done to foster their success in STEM is timely as more and more public attention has begun to focus on the circumstances of first-generation college students (see for example the May 14, 2015 issue of Chronicle of Higher Education). And while research shows that earning a bachelor's degree in STEM doubles the earning potential of first-generation students as compared to those only completing high school (The Hamilton Project, 2016), data regarding first-generation college students shows that they are disproportionately at risk of dropping out of college without completing a degree (Greenwald, 2012; Lohfink & Paulsen, 2005).

This Book

This book will present evidence that an important reason for the lack of success of first-generation students in STEM is their lack of familiarity with the landscape of higher education and their lack of savvy about how to study for college level science courses. Furthermore, this book will describe (for faculty, graduate students, and undergraduates) how first-generation college students can easily learn or be taught to develop more effective study skills that can enable them to be successful in introductory, so called "gatekeeping" STEM courses. Specifically, this book will discuss how to teach students to become "self-regulated" learners, learners who are strategic, who manage their time effectively, who monitor their performance, and who utilize feedback to change their study habits and improve their performance (Zimmerman & Schunk, 2001).

References

Chang, M. J., Eagan, M. K., Lin, M. H., & Hurtado, S. (2011). Considering the impact of racial stigmas and science identity: Persistence among biomedical and behavioral science aspirants. *Journal of Higher Education, 82*(5), 564–596.

Davis, J. (2010). *The first-generation student experience.* Sterling, VA: Stylus.

Eberlein, T., Kampmeier, J. A., Minderhout, V., Moog, R. S., Platt, T., Varma-Nelson, P., & White, H. B. (2008). Pedagogies of engagement in science: A comparison of PBL, POGIL, and PLTL. *Biochemistry and Molecular Biology Education, 38*(4), 262–273.

Freeman, S., Eddy, S. L., McDonough, M., Smith, M. K., Okoroafor, N., & Jordt, H. (2014). Active learning increases student performance in science, engineering, and mathematics. *Proceedings of the National Academy of Sciences, 111*(23), 8410–8415.

Gasiewski, J. A., Eagan, M. K., Garcia, G. A., Hurtado, S., & Chang, M. J. (2012). From gatekeeping to engagement: A multicontextual, mixed method study of student academic engagement in introductory STEM courses. *Research in Higher Education, 53*, 229–261.

Greenwald, R. (2012). Think of first-generation students as pioneers, not problems. *The Chronicle of Higher Education, 59*, 12.

Hunter, A. B. (2016, April). *Factors contributing to undergraduate switching from stem majors: Has anything changed in the last two decades?* Paper presented at the Annual Conference of the American Educational Research Association, Washington, DC.

Hurtado, S., Newman, C. B., Tran, M. C., & Chang, M. J. (2010). Improving the rate of success for underrepresented racial minorities in STEM fields: Insights from a national project. *New Directions for Institutional Research, 148*, 5–15.

Kober, L. (Ed.) (2015). *Reaching students: What research says about effective instruction in undergraduate science and engineering.* Washington, DC: The National Academies Press.

Lohfink, M. M., & Paulsen, M. B. (2005). Comparing the determinants of persistence for first-generation and continuing-generation students. *Journal of College Student Development, 46*(4), 409–428.

Malcom, S., & Feder, M. (Eds.). (2016). *Barriers and opportunities for 2-year and 4-year STEM degrees: Systemic change to support students' diverse pathways.* Washington, DC: National Academies Press.

Russell, M. L., & Atwater, M. M. (2005). Traveling the road to success: A discourse on persistence throughout the science pipeline with African American students at a predominantly White institution. *Journal of Research in Science Teaching, 42*(6), 691–715.

Seymour, E., & Hewitt, N. M. (1997). *Talking about leaving: Why undergraduates leave the sciences.* Boulder, CO: Westview Press.

The Hamilton Project. (2016). *Career earnings by college major.* Retrieved from http://www.hamiltonproject.org/charts/career_earnings_by_college_major/

Treisman, P. U. (1992). Studying students studying calculus: A look at the lives of minority mathematics students in college. *The College Mathematics Journal, 23*(5), 362–372.

Zimmerman, B., & Schunk, D. H. (Eds.). (2001). *Self-regulated learning and academic achievement: Theoretical perspectives* (2nd ed.). Mahwah, NJ: Erlbaum.

2

First-Generation College Students

Leigh McCallen

Access to higher education in the United States has broadened signifi-cantly in the past few decades, with more than a third of the population of adults over age 25 now holding a bachelor's degree (Snyder, de Brey, & Dillow, 2016). Twinned with increased access to postsecondary education is a decrease in the total share of individuals with parents who have not attended college: Among high school sophomores, 77% had parents with-out postsecondary education in 1980 versus 62% in 2002 (Cahalan, Ingels, Burns, Planty, & Daniel, 2006), and among college enrollees, the propor-tion of students whose parents have not attended college declined 37% to 33% between 1999–2000 and 2011–2012 (Skomsvold, 2015). Despite higher education becoming more accessible, these students—referred to as "first-generation college students"—still comprise over a third of total college-goers nationwide (Skomsvold, 2015) and more than 20% of the 7 million undergraduates attending 4-year institutions (Pappano, 2015).

Teaching STEM to First Generation College Students, pages 7–19
Copyright © 2019 by Information Age Publishing

Defining First-Generation College Student Status

In the broadest sense, status as a first-generation college student refers to individuals whose parents have not participated in postsecondary education and have no education beyond high school. This is the definition used by, for example, the National Center for Education Statistics and the Higher Education Research Institute (Cataldi, Bennett, & Chen, 2018; Pryor et al., 2006). However, narrower parameters of first-generation status abound: The Department of Education's legislative definition specifically refers to individuals with parents who do not have a bachelor's degree, and a recent report from the Institute for Higher Education Policy calls for a definition that excludes individuals with parents who have an associate's degree (Sharpe, 2017). The Pell Institute for the Study of Opportunity in Higher Education also defines first-generation status as those individuals with parents who do not have bachelor's degrees (Engle & Tinto, 2008; Engle, Bermeo, & O'Brien, 2006).

First-Generation Students and Barriers to College Success

The majority of the research literature concerning the outcomes of first-generation students employ the more general definition of being the first person in one's family to attend college, with the lion's share of findings indicating that salient barriers to college success exist for these students. First-generation students are more likely to leave college and less likely to earn a degree as compared to peers with college-educated parents (Chen, 2005). According to a recent report from the National Center for Educational Statistics (Cataldi et al., 2018), first-generation students are more likely to enter higher education at public 2-year colleges and less likely to enter at public 4-year colleges as compared to continuing-generation peers (those whose parents graduated college or have experience with college). These disparities continue to follow first-generation students through their college careers: Three years after college enrollment, first-generation students are more likely to leave without a credential, and 6 years after postsecondary entry, fewer first-generation students earn a credential or remain enrolled compared to continuing-generation peers, regardless of whether they began at a public 2-year college, private 4-year college, or public 4-year college (Cataldi et al., 2018). In fact, nearly 90% of first-generation college students fail to graduate within 6 years of enrollment (Lohfink & Paulsen, 2005). However, if first-generation students are able to navigate college successfully to completion, they accrue significant benefits: Among bachelor's degree recipients, there are no significant differences between first-generation and continuing-generation students in terms of labor market outcomes 4 years after graduation as measured by full-time employment status or median salary (Cataldi et al., 2018).

Differences in the outcomes of first-generation students are attributable in part to the fact that their status overlaps with other social and demographic factors shown to independently limit college success, such as being older than 24 years, working full-time, delaying postsecondary enrollment, attending college part-time, being financially independent, and/or supporting dependents (Greene, Marti, & McClenney, 2008; Engle, 2007; Lohfink & Paulsen, 2005). Possessing any one of these characteristics, in addition to being first-generation in college, Black or Latino/a, and/or low-income, has been shown to greatly increase the chance of dropping out without a credential, and for those contending with two or more characteristics, only 25% will eventually earn a degree (Adelman, 2005).

Low-income status is an especially important intersecting factor. First-generation students are more likely to come from low-income backgrounds, with 27% coming from households making $20,000 or less and 50% from households making between $20,001 and $50,000, as compared to 6% and 23% of continuing-generation students, respectively (Redford & Hoyer, 2017). In absolute terms, 4.5 million low-income first-generation students are enrolled in postsecondary education, representing 24% of the undergraduate population (Engle & Tinto, 2008). Additionally, more first-generation students (54%) than continuing-generation students (45%) cite not being able to afford college as a reason for leaving without a credential (Redford & Hoyer, 2017), and loans make up a greater proportion of financial aid packages for low-income first-generation students (Engle & Tinto, 2008). Furthermore, they are prone to experience greater isolation and marginalization in college as compared to peers with college educated parents (Jehangir, 2010).

Theoretical Approaches to Understanding the College Success of First-Generation Students

Current knowledge about the mechanisms driving college success outcomes of first-generation college students has a significant base in sociology, specifically in terms theorizing about the effect of social capital (resources) on social stratification and the role of engagement in college and campus life. Traditionally, each of these approaches has focused on what first-generation students lack in terms of social resources and engagement as a way to explain differential outcomes as compared to continuing-generation or otherwise more advantaged students. However, contemporary researchers are now articulating theories that can frame first-generation student success in ways that challenge the traditional deficit-laden framework. This chapter will provide an overview of traditional and contemporary theorizing about first-generation college students in terms of social capital and college engagement, ending with a discussion of integrated and contextual theory that attempts to bring together multiple strands of research to provide a more comprehensive lens through which to understand the college trajectories of first-generation students.

Social and Cultural Capital

Traditional Framing

Beyond coping with an overlapping constellation of social and demographic factors that independently limit retention and performance, researchers have found first-generation students face specific challenges related to limited family support (Engle, 2007; Pascarella, Wolniak, Pierson, & Terenzini, 2003). First-generation students must cope with a lack of knowledge about higher education, and even as socioeconomic status increases, their disadvantage in terms of college completion is not eliminated due to variations in parental knowledge, resources, college-specific experiences, and family stressors (Wilbur & Roscigno, 2016). As Stephens, Hamedani, and Destin (2014) put it, first-generation college students "struggle to navigate the middle-class culture of higher education, learn the 'rules of the game,' and take advantage of college resources" (p. 944). They go on to say, however, that "because U.S. colleges and universities seldom acknowledge how social class can affect students' educational experiences, many first-generation students lack insight about why they are struggling and do not understand how students 'like them' can improve." (p. 944) For example, first-generation students may not even know that resources such as tutoring or faculty office hours are available and can be useful—even integral—to their success, while continuing-generation students arrive on campus with a greater degree of knowledge about the types of supports that exist and how to harness them to their advantage (Nichols & Islas, 2016; Winograd & Rust, 2014; Collier & Morgan, 2008).

These differences have been widely attributed to disparities in access to social and cultural capital between first-generation students and their middle or upper-class peers with college-educated parents. Social capital is defined as the relational connections and trust within a social organization or group maintained through family, peer, and other social networks that influence individuals' capacity to navigate institutions effectively, including access to economic resources (Coleman, 1988; Putnam, 1995; Yosso, 2005). Cultural capital refers to the noneconomic norms and information channels that are transmitted through social networks that enable class mobility (Bourdieu, 1986). It is theorized that the social and cultural capital of the upper classes is more valuable within the hierarchy of society, and thus contributes to reproducing the prevailing structure through intergenerational transmission within families and social institutions, such as universities (Bourdieu, 1986; Lohfink & Paulsen, 2005).

Having attended college, parents of continuing-generation college students can draw on first-hand knowledge to help navigate their children through the process of applying to college, transitioning to college, and succeeding once there (Jehangir, 2010). It is typically theorized that first-generation students' deficiency in having access to social and cultural capital underlies their struggle with institutional expectations and their identity as a college student (Ward, Siegel, & Davenport, 2010) because college-related cultural capital provides a significant, ingrained orientation to the college experience (Davis, 2010).

Contemporary Framing

Yosso (2005) argues against the traditional framing of social capital, taking issue with underlying assumption that the academic and social outcomes of people of color and other underrepresented groups are rooted in these social groups' lack of the cultural resources necessary for social mobility. Drawing on research in education using critical race theory, Yosso summarizes six forms of cultural capital nurtured within marginalized communities that promote social mobility: aspirational capital, defined as the capacity to maintain optimism and motivation in the face of real and perceived barriers; linguistic capital, the skills developed through experiences in more than one language; familial capital, the cultural knowledge of families; social capital, the networks of people who provide instrumental and emotional support; navigational capital, or skills of moving through and coping with social institutions; and resistant capital, the attitudes developed through oppositional behavior to challenge inequality. For example, Gofen (2009) found first-generation students' family capital to be a significant and multifaceted success-promoting influence rather than a limiting factor, specifically socially and culturally situated family psychosocial resources such as habits, educational priorities, emotional support, belief systems, and educational values.

Other research in the vein of social and cultural capital has articulated how resources can be harnessed to increase the social mobility of marginalized groups, including first-generation students. Stanton-Salazar (1997) identifies the importance of institutional agents, "those individuals who have the capacity and commitment to transmit directly, or negotiate the transmission of, institutional resources and opportunities" (p. 6), individuals such as teachers or counselors who convey cultural capital found to impact racial/ethnic minority students' educational outcomes. Specifically, institutional agents seek to elevate the opportunities of marginalized students by providing funds of knowledge (such as knowledge of institutionally sanctioned discourses and knowledge of how to navigate organizational bureaucracies), bridging to institutional gatekeepers, engaging in advocacy and personalized intervention, role modeling, and providing emotional and moral support.

Dowd, Pak, and Bensimon (2013) examined the role of institutional agents in promoting the transfer of low-income and students of color from a community college to selective 4-year colleges, finding that 4-year college faculty members were instrumental in providing a sense of psychological security and validation through their relationship with low-status students, which in turn supported the formation of an "elite" academic identity. Similarly, McCallen (2016) found that what most differentiated academically successful and nonsuccessful first-generation students at urban public 4-year colleges was access to the institutional support of college faculty: Significant college faculty played a pivotal role by conveying not only encouragement, but also navigational and intellectual institutional resources that together facilitated students' access to academic support, sense of institutional belonging, and solidified their academic/career identities.

College Engagement

Traditional Framing

Student engagement in college is a multifaceted construct and results from the interaction between students' individual characteristics that facilitate learning and the institutional conditions that facilitate engagement in educational activities, such as effective faculty teaching practices and features of the college environment (Habley, Bloom, & Robbins, 2012; Umbach & Wawrzynski, 2005). The National Survey of Student Engagement (NSSE) was developed to build on robust evidence in the field of higher education that increasing the engagement of students, especially for those whom engagement is known to be problematic and result in negative outcomes, is worth systemic investigation across all types of institutions (Harper & Quaye, 2015). The instrument has been administered to over four million undergraduates at more than 1,500 institutions since 2000, and measures a set of engagement indicators, including academic challenge, learning with peers, experiences with faculty, and high-impact educational practices (Harper & Quaye, 2015). Results from the NSSE and its community college counterpart, the Community College Survey of Student Engagement (CCSSE), consistently indicate a positive relationship between engagement and positive academic outcomes in college, such as GPA and persistence (Nakamaru, 2012).

Engagement has been demonstrated to be an important process for educationally at-risk students in terms of persistence and retention in higher education. In the literature, engagement is defined as representing (a) the effort students commit to educationally purposeful activities (Greene et al., 2008; Nakamaru, 2012), and (b) how higher education institutions transmit resources, organize learning opportunities, and provide support services to "induce" students to participate in activities leading to positive outcomes such as persistence and graduation (Kuh, Kinzie, Buckley, Bridges, & Hayek, 2007). In comparison to peers with college-education parents, first-generation college students are less engaged overall, perceive the college environment as less supportive, and reported making less progress in their learning and intellectual development; interestingly, the authors attribute first-generation students' lack of engagement to their limited knowledge of the importance of engagement itself (Pike & Kuh, 2005).

Greene and colleagues (2008) suggest the link between engagement and academic outcomes may be a crucial factor in reducing achievement gaps in higher education, particularly among students attending community colleges or remedial programs. For example, Nakamaru (2012) showed a strong, positive relationship between urban community college students' level of engagement with technology (a class wiki) in a remedial English as a second language (ESL) class, and students' success in exiting remediation by passing a writing exam.

Contemporary Framing
Engagement in educationally purposeful activities has been shown to relate positively to academic outcomes in terms of retention, suggesting that institutions should develop ways to engage students in effective educational practices, especially those who are first-generation in college and/or from low-income backgrounds. There also appears to be a "differential return" on engagement, as at-risk students may benefit more from college engagement than their high-achieving peers in terms of positive academic outcomes (Kuh, Cruce, Shoup, Kinzie, & Gonyea, 2008).

However, the traditional framing of engagement has tended to place more emphasis on the effort students devote to academic and social activities as opposed to how institutions can better structure resources and supports to help underrepresented students. For example, Kezar, Walpole, and Perna (2015) argue institutions have not paid specific and sufficient attention to the engagement of low-income students (many of whom are first-generation in college), and offer recommendations for engaging this population, such as ensuring basic financial needs are met, focus on engagement in the classroom (instead of relying on activities outside the classroom), offer encouragement and support to students who work, and create experiences that are culturally relevant in terms of students' racial and social class backgrounds. As Engle and Tinto argue (2008), institutions must also provide professional development for faculty and staff to help them acquire pedagogical skills that can be used with at-risk low-income and first-generation students.

Integrated Contextual Models

Perna and Thomas (2008) propose that integrated approaches addressing multiple influences and layers of context are necessary to generate new insights about the success of underrepresented students with the goal of better informing the work of policymakers and practitioners interested in reducing inequality in higher education. In the authors' theoretical model, internal context is conceptualized as the core of student success, determined by the attitudes, motivations, and behaviors of individual students. Just outside the core of student success is family context, which the authors describe as the ways families contribute to children's experiences with the intent of promoting success. At the layer of school context, the authors describe how institutional economic factors and financial aid impact college persistence and completion. At the most distal layer of the model, social, economic, and policy factors highlight how external forces influence student success directly and indirectly through proximal contexts, for example, how the public financing of K–12 schools influence college choice and success.

Like Perna and Thomas' model (2008), Bronfenbrenner's ecological systems theory (1979) attends to contexts that reciprocally affect individual development within and across settings. This theoretical framework has been identified as useful for elucidating individual, social, and institutional factors that foster positive educational outcomes among students confronting barriers to success in higher education (Gayles, 2005; Morales, 2012), further aiming to promote research on educational equity in socially meaningful ways by understanding the barriers underrepresented students cope with and the conditions under which some students do well (Morales &

Trotman, 2004; O'Dougherty-Wright, Masten, & Narayan, 2013). Demetriou, Meece, Eaker-Rich, and Powell (2017) applied ecological theory to consider multiple domains impacting successful first-generation college students' positive trajectories, finding that these students actively sought out activities, learning experiences, and particularly mentoring relationships in increasingly complex ways that led to positive adaptation to the college environment, and ultimately, student retention.

Morales and Trotman (2011) also applied an ecological model of student resilience, positing that resilience is a process of overcoming challenges that is dependent on features of the social environment rather than an innate personality characteristic. To this end, the authors investigated multiple layers of influence to identify two clusters of protective factors first-generation students of color draw on to become successful in higher education.

The first cluster, called "It's Okay to be Smart: Skillful Mentoring for Future Success," involved the interplay of the following dispositional and environmental factors: desire for social mobility (to "class jump"), caring school personnel (K–12 and college), sense of obligation to one's race/ethnicity, and a strong future orientation. This cluster, Morales notes, reflected the ways that supportive teachers and other school/college personnel serve as academic role models for at-risk students, and help students overcome specific stressors related to minority background and class status. In the study, 66% of students interviewed reported a clear interplay between these four protective factors.

Cluster 2, called "Pride, Debt, Effort, and Success: Becoming Someone" was reported by 70% percent of students as playing a crucial role in their academic success. This cluster consisted of seven factors, primarily in the dispositional and familial domains: strong work ethic, persistence, high self-esteem, internal locus of control, attendance at "non-neighborhood" school (K–12), high parental expectations supported by words and actions, and a mother modeling a strong work ethic. As discussed by Morales, Cluster 2 reflected the interplay of core resilience-promoting dispositional factors and the strong influence of parental modeling, involvement, and cultural values.

In another study using a similar approach, Morales (2012) conducted a prospective longitudinal, qualitative study over the course of first-generation college students' initial semesters, comparing interviews with students to elucidate school environment factors and personal dispositions that contributed to eventual academic success or failure. Key attributes of successful students in the study included a willingness to seek help from a variety of resources, acknowledgment of potential academic issues or deficits, and students' self-imposed structure for studying during free time, particularly at the beginning of the semester. Morales discusses practical implications of these findings, such as the importance of supporting students towards early success at the beginning of the semester by providing detailed feedback on assignments so that students can evaluate where they stand and where they need to be.

Framing of First-Generation Students in This Book

We present a number of sociological perspectives through which to view first-generation students with the hope that readers of this book will come to regard first-generation students not in terms of their deficiencies, but in terms of the breadth, depth, and richness of experience that they bring to the classroom. Following from contemporary theoretical approaches reviewed here, this book takes a purposeful stance towards framing first-generation students from a strengths-based perspective that acknowledges these students must be provided with the proper tools to help them succeed—and that institutions of higher education, particularly faculty, have an important role to play in helping first-generation students succeed. More specifically, this book focuses on how to help first-generation students acquire and improve upon their self-regulated learning skills, abilities which are a type of cultural capital necessary to succeeding in college that continuing-generation students possess more of.

References

Adelman C. (2005). *Moving into town and moving on: The community college in the lives of traditional–age students.* Washington, DC: U.S. Department of Education.

Bourdieu, P. (1986). The forms of capital. In A. R. Sadovnik (Ed.), *Sociology of education: A critical reader* (pp. 83–96). New York, NY: Routledge.

Bronfenbrenner, U. (1979). *The ecology of human development: Experiments by nature and design.* Cambridge, MA: Harvard University Press.

Cahalan, M. W., Ingels, S. J., Burns, L. J., Planty, M., & Daniel, B. (2006). *United States high school sophomores: A twenty-two year comparison, 1980–2002.* (NCES 2006-327). U.S. Department of Education. Washington, DC: National Center for Education Statistics.

Cataldi, E. F., Bennett, C., & Chen, X. (2018). *First-generation students: College access, persistence, and postbachelor's outcomes.* (NCES 2018-421). Washington, DC: National Center for Education Statistics, U.S. Department of Education.

Chen, X. (2005). *First generation students in postsecondary education.* U.S. Department of Education, National Center for Education Statistics. Washington, DC: U.S. Government Printing Office.

Coleman, J.S. (1988). Social capital in the creation of human capital. *American Journal of Sociology, 94* (Issue Supplement), S95–S120.

Collier, P. J., & Morgan, D. L. (2008). "Is that paper really due today?": Differences in first-generation and traditional college students' understandings of faculty expectations. *Higher Education, 55,* 425–446.

Davis, J. (2010). *The first generation student experience.* Sterling, VA: Stylus.

Demetriou, C., Meece, J., Eaker-Rich, D., & Powell, C. (2017). The activities, roles, and relationships of successful first-generation college students. *Journal of College Student Development, 58*(1), 19–36.

Dowd, A. C., Pak, J. H., & Bensimon, E. M. (2013). The role of institutional agents in promoting transfer access. *Education Policy Analysis Archives, 21*(15), 1–44.

Engle, J. (2007). Postsecondary access and success for first-generation college students. *American Academic, 3,* 25–48.

Engle, J., Bermeo, A., & O'Brien, C. (2006). *Straight for the source: What works for first-generation college students.* Washington, DC: The Pell Institute for the Study of Opportunity in Higher Education.

Engle, J., & Tinto, V. (2008). *Moving beyond access: College success for low-income, first-generation students.* Washington, DC: The Pell Institute for the Study of Opportunity in Higher Education.

Gayles, J. (2005). Playing the game and paying the price: Academic resilience among three high-achieving African-American males. *Anthropology and Education Quarterly, 36*(3), 250–264.

Greene, T. G., Marti, N., & McClenney, K. (2008). The effort-outcome gap: Differences for African-American and Hispanic community college students in student engagement and academic achievement. *The Journal of Higher Education, 79*(5), 513–539.

Gofen, A. (2009). Family capital: How first-generation higher education students break the intergenerational cycle. *Family Relations, 58*(1), 104–120. doi:10.1111/j.1741-3729.2008.00538

Habley, W. R., Bloom, J. R., & Robbins, S. (2012). *Increasing persistence: Research-based strategies for college student success.* San Francisco, CA: Jossey-Bass.

Harper, S. R., & Quaye, S. J. (2015). Making engagement equitable for students in U.S. higher education. In J. Quaye & S. R. Harper (Eds.), *Student engagement in higher education* (pp. 1–14). New York, NY: Routledge.

Jehangir, R. R. (2010). *Higher education and first-generation students: Cultivating community, voice, and place for the new majority.* New York, NY: Palgrave MacMillan.

Kezar, A. J., Walpole, M., & Perna, L. W. (2015). Engaging low-income students. In J. Quaye & S. R. Harper (Eds.), *Student engagement in higher education* (pp. 237–256). New York, NY: Routledge.

Kuh, G. D., Cruce, T. M., Shoup, R., Kinzie, J., & Gonyea, R. M. (2008). Unmasking the effects of student engagement on first-year college grades and persistence. *The Journal of Higher Education, 79*(5), 540–563.

Kuh, G. D., Kinzie, J., Buckley, J., Bridges, B., & Hayek, J. C. (2007). Piecing together the student success puzzle: Research, propositions, and recommendations. *ASHE Higher Education Report, 32*(5). San Francisco, CA: Jossey-Bass.

Lohfink, M. M., & Paulsen, M. B. (2005). Comparing the determinants of persistence for first-generation and continuing-generation students. *Journal of College Student Development, 46*(4), 409–428.

McCallen, L. S. (2016). The critical social ecology of student success in higher education: A transformative mixed methods study of undergraduates' experiences and outcomes at the City University of New York (Doctoral dissertation). Retrieved from ProQuest. (1553)

Morales, E. E. (2012). Navigating new worlds: A real-time look at how successful and non-successful first-generation college students negotiate their first semesters. *International Journal of Higher Education, 1*(1), 90–100.

Morales, E. E., & Trotman, F. K. (2004). *Promoting academic resilience in multicultural America: Factors affecting student success.* New York, NY: Peter Lang.

Morales, E. E., & Trotman, F. K. (2011). *A focus on hope: Fifty resilient students speak.* Lanham, MD: University Press of America.

Nakamaru, S. (2012). Investment and return: Wiki engagement in a "remedial" ESL writing course. *Journal of Research on Technology in Education, 44*(4), 273–291.

Nichols, L., & Islas, A. (2016). Pushing and pulling emerging adults through college: College generational status and the influence of parents and others in the first year. *Journal of Adolescent Research, 31*(1), 59–95.

O'Dougherty-Wright, M., Masten, A.S., & Narayan (2013). Resilience processes in development: Four waves of research on positive adaptation in the context of adversity. In S. Goldstein & R. B. Brooks (Eds.), *Handbook of resilience in children* (pp. 15–39). New York, NY: Springer.

Pascarella, E. T., Wolniak, G. C., Pierson, C. T., & Terenzini, P. T. (2003). Experiences and outcomes of first-generation students in community colleges. *Journal of College Student Development, 44*(3), 420–429.

Pappano, L. (2015). First-generation students unite. *The New York Times,* Retrieved from https://www.nytimes.com/2015/04/12/education/edlife/first-generation-students-unite.html

Perna, L. W., & Thomas, S. L. (2008). Theoretical perspectives on student success: Understanding the contributions of the disciplines. *ASHE Higher Education Report, 34*(1), 1–63.

Pike, G. R., & Kuh, G. D. (2005). First- and second-generation college students: A comparison of their engagement and intellectual development. *The Journal of Higher Education, 76*(3), 276–300.

Pryor, J. H., Hurtado, S., Saenz, V. B., Korn, J. S., Santos, J. L., & Korn, W. S. (2006). *The American freshman: National norms for fall 2006, Cooperative Institutional Research Program.* Higher Education Research Institute Graduate School of Education & Information Studies, University of California, Los Angeles.

Putnam, R. D. (1995). Bowling alone: America's declining social capital. *Journal of Democracy, 6*(1), 65–78.

Redford, J., & Hoyer, K. M. (2017). *First-generation and continuing-generation college students: A comparison of high school and postsecondary experiences* (NCES 2018-009). U.S. Department of Education. Washington, DC: National Center for Education Statistics.

Sharpe, R. (2017). Are you first-gen? Depends on who's asking. *The New York Times.* Retrieved from https://www.nytimes.com/2017/11/03/education/edlife/first-generation-college-admissions.html

Skomsvold, P. (2015). *Web tables—Profile of undergraduate students: 2011–12* (NCES 2015-167). Washington, DC: U.S. Department of Education.

Snyder, T. D., de Brey, C., & Dillow, S. A. (2016). *Digest of education statistics, 2015* (NCES 2016-014). Washington, DC: U.S. Department of Education.

Stanton-Salazar, R. D. (1997). A social capital framework for understanding the socialization of racial minority children and youth. *Harvard Educational Review, 67*(1), 1–40.

Stephens, N. M., Hamedani, M. G., & Destin, M. (2014). Closing the social-class achievement gap: A difference-education intervention improves first generation students' academic performance and all students' college transition. *Psychological Science, 25*(4), 943–953.

Umbach, P. D., & Wawrzynski, M. R. (2005). Faculty do matter: The role of college faculty in student learning and engagement. *Research in Higher Education, 46*(2), 153–184.

Ward, L., Siegel, M. J., & Davenport, Z. (2010). First generation college students: Understanding and improving the experience from recruitment to commencement. San Francisco, CA: Jossey-Bass.

Wilbur, T. G., & Roscigno, V. J. (2016). First-generation disadvantage and college enrollment/completion. *Socius: Sociological Research for a Dynamic World, 2*, 1–11.

Winograd, G., & Rust, J. P. (2014). Stigma, awareness of support services, and academic help-seeking among historically underrepresented first-year college students. *The Learning Assistance Review, 19*(2), 17–41.

Yosso, T. J. (2005). Whose culture has capital? A critical race theory discussion of community cultural wealth. *Race Ethnicity and Education. 8*(1), 69–91.

3

Success in STEM

It's Not Just About Effort, Intelligence, or High School Background

I used to believe wholeheartedly that there are smart students and slow students . . .
I now know that there are students who have an arsenal of strategies
at their disposal and there are students who don't.

—Dr. Sandra McGuire (McGuire & McGuire, 2015, p. 65)

Study Skills Matter

Are Some Students Just Better at STEM?

Dr. Sandra McGuire was a chemistry professor at Cornell University and Louisiana State University. She went on to direct the Center for Academic Success at LSU and eventually served as the assistant vice chancellor there. Like many science faculty (including myself), she began her teaching career believing that there were smarter students, students who were more gifted in science, and students who would always struggle to succeed science.

My story is not so different. After having taught chemistry for 20 years (mostly laboratory courses for majors and lecture courses for nonscience

Teaching STEM to First Generation College Students, pages 21–30
Copyright © 2019 by Information Age Publishing
21

majors) at a small, private, liberal arts college, I relocated to CUNY-Brooklyn College, a large, urban, public university. I was hired by Brooklyn College to teach organic chemistry lecture courses and was specifically assigned the task of helping improve student success rates, which I was told were only about 50%.

I arrived at Brooklyn College expecting that students would find organic chemistry difficult and would struggle, either because of the subject's difficulty, or because of its daunting reputation (Grove, Hershberger, & Bretz, 2008) which might interfere with student confidence and motivation.

I was quite surprised, however, to discover, that student success in Organic Chemistry I at Brooklyn College was strongly tied to and predicted by students' prior performance in general Chemistry II. The simple Pearson correlation coefficient for the relationship between performance in Organic Chemistry I and General Chemistry II was 0.70 ($p < 0.001$) (Horowitz, Rabin, & Brodale, 2013). In other words, just under 50% of the variability in Organic I grades was explained by prior performance in General Chemistry II.

Knowing that General Chemistry II focuses primarily on topics like thermodynamics and kinetics, which are not particularly important for organic chemistry, this finding was quite surprising and perplexing. In fact, General Chemistry II is not even a required prerequisite for Organic Chemistry I at some institutions (Ege, Coppola, & Lawton, 1997; Malinak et al., 2014).

Help Seeking Matters

Further research, however, helped clarify and explain. Results demonstrated that when *controlling for prior performance in General Chemistry II*, academic help seeking behavior explained 10% of variability in performance in Organic Chemistry I ($p < 0.001$) while prior performance in General Chemistry II now only explained 31% ($p < 0.001$; Horowitz, 2011). To state this in a simpler way, students who started out academically weaker (who had poor grades in General Chemistry II), but who engaged in help seeking behaviors (such as attending supplementary problem sessions and office hours), were able to overcome the odds and pass Organic Chemistry I. But their counterparts, who had also done poorly in General Chemistry II, but who did not engage in help seeking while taking Organic Chemistry I, were not able to successfully get through the Organic Chemistry I course. Thus, the strong correlation between performance in General Chemistry II and Organic Chemistry I was in part due to a skill set possessed by successful science students—*that of knowing how and when to seek help* (Horowitz et al., 2013).

Similarly researchers at another public university found that study behaviors of organic students predicted performance in organic chemistry and outweighed other predictive factors such as prior GPA (Szu et al., 2011). Additionally, they found that successful organic chemistry students started studying and engaging in help seeking behaviors earlier in the semester (Szu et al., 2011).

You might be thinking that measures such as prior GPA or performance in a prerequisite course like general chemistry may measure factors such as intelligence, aptitude towards science, or level of high school science preparation, and that therefore it makes good sense as to why such factors predict organic chemistry grades. However, it was both fascinating and heartening to myself and my co-authors to observe and find literature support for the fact that student behaviors that could be loosely grouped together under the general heading of "study skills" could enable "at-risk" students to survive and succeed in a challenging course like organic chemistry. Furthermore when we started to dig deeper, we began to understand and realize that success in STEM at the college level is often in reality about the development of sophisticated study skills and not necessarily about aptitude (e.g., how smart you are, or how good at science you are).

Self-Regulated Learning

Help seeking and other types of study skills and behaviors are part of a broader psychological construct known as self-regulated learning (Zimmerman, 1990). Zimmerman (1990) defines self-regulated learning (SRL) as a process in which individuals monitor their own learning processes, employ strategies to meet their goals, and modify their behaviors in response to feedback. A number of studies have demonstrated that engaging in help seeking and/or other forms of SRL behaviors correlates with better course grades in college level science courses (Bunce et al., 2017; Dibenedetto, 2010; Dibenedetto & Bembenutty, 2013; Karabenick, 2003; Nandagopal & Ericsson, 2012; Sebesta & Bray-Speth, 2017; VanderStoep, Pintrich, & Fagerlin, 1996).

Study Skill Training Improves Stem Performance

A recent study (Hudesman et al., 2014) took things one step further, demonstrating that college level students can be trained to adopt SRL behaviors and thereby improve their course performance. Specifically, students described as urban, "at risk," and attending a vocational college were trained to review and correct their own quizzes as part of the interventional study embedded within a developmental mathematics course. As a result of this intervention, treatment students outperformed control students by 25% on a national mathematics examination.

Comparable results were found in general chemistry (Cook, Kennedy, & McGuire, 2013) where researchers demonstrated that students who attended a 50-minute training regarding study and learning strategies outperformed a control group by one full letter grade! Faculty across STEM disciplines have implemented training programs like this general chemistry one and/or have utilized a pedagogical tool known as an exam wrapper (where students assess themselves and reflect on their own learning and study habits [Lovett, 2013]) to train students to improve their SRL, with corresponding improvements in student performance in subjects such as biology, mathematics, and statistics (Azevedo & Cromley, 2004; Bernacki, Vosicka, & Utz, 2016; Chen, Chaves, Ong, & Gunderson, 2017; Olszewski, 2016). Others have utilized incentive methods to train students to correct and learn from their exam errors (similar to the Hudesman et al., 2014 study) and have likewise shown improvements in student performance in physics and biology courses (Brown, Mason, & Singh, 2016; Henderson & Harper, 2009; Rozell, Johnson, Sexten, & Rhodes, 2017).

First-Generation Students and Self-Regulated Learning

First-Generation Students

But what does the connection between SRL (or well developed study skills) and performance in STEM have to do with the topic of this book, which is to help first-generation college students do better in STEM? And what do I mean when I use the term first-generation college student?

Various definitions of what it means to be a first-generation college student can be found in the literature (Davis, 2010; Jehangir, 2010; Ward, Siegel, & Davenport, 2012). Some define a first-generation college student as one who is the first in his or her family to earn any college credit. Others define a first-generation student as the first in his or her family to earn a degree, whether this degree be a certificate, an associate's degree or a bachelor's degree. This book focuses on the success of first-generation students in the most comprehensive sense, students who are the first-generation to complete a bachelor's degree, because research shows that students who fit either definition are at higher risk of leaving college without earning a degree or credential (Cataldi, Bennett, & Chen 2018).

Self-Regulated Learning

Research shows that first-generation students often suffer from a lack of knowledge about higher education and that this impedes their success in college (Dennis, Phinney, & Chuateco, 2005; Stephens, Hamedani, & Destin, 2014; Ward et al., 2012). When it comes to success in STEM, of concern is that fact that the insight that first-generation students often lack is not just a general one, but is a specific lack of knowledge about how to study effectively—how to monitor one's performance and adjust study behaviors when necessary (Davis, 2010; Jehangir, 2010; Thorne, 2013). The repertoire of study behaviors of first-generation students may be more limited (Weinstein & Van Mater Stone, 1993) than that of continuing generation students and first-generation students may not feel comfortable seeking help when they need it (Davis, 2010; Kim & Sax, 2009; Newman, 2000; Stephens, Markus, & Phillips, 2014).

Working in a diverse institution like CUNY-Brooklyn College where approximately 45% of students do not have a parent with a college degree (CUNY, 2014), I have observed over and over how my continuing generation science students seem to possess an entirely different repertoire of academic skills and behaviors from my first-generation students. But, research demonstrates that first-generation STEM students benefit from training in SRL. Research shows that when first-generation STEM students are trained in "learning how to learn," their course performance improves significantly (Bernacki, Vosicka, & Utz, 2016; Bernacki, Vosicka, & Utz, 2017).

So Why This Book?

Many of us who have earned graduate degrees in STEM and/or who teach STEM in higher education sometimes wonder what is up with our students. Why do they struggle so much? Why is it so hard for them to succeed? Sometimes we find ourselves wondering if they are just not cut out for science or math. Maybe they are just not smart enough or just don't put in the effort? We likely grew up "good" at math or science. And we certainly had to bust our butts to succeed in school and achieve the successes and accomplishments we have obtained. So it's natural to expect the same level of achievement from our students. But we sometimes forget that we are in effect professional students—we have honed the skills of being expert students, expert scholars, and expert academics (Ertmer & Newby, 1996; Weinstein & Van Mater Stone, 1993). It is therefore often hard to remember how, where, or when we acquired these skills.

Furthermore, we may forget or be only subliminally aware of the advantages that we might have had growing up. Research shows that the majority of university faculty have had the benefit of growing up in a family where one of more of their parents had been to college (Hoffer et al., 2001; Kniffin, 2007). So like "fish in water" (Reay, David, & Ball, 2005), our transition to the social milieu of college may have been a rather smooth one.

In my case, both my parents had graduate degrees. My father worked as a civil engineer and my mother was a high school social studies teacher. While my mother directly and indirectly taught me how to be a successful student (she "supervised" as I sat evenings at our kitchen table doing my homework), my father was the one who encouraged my interest in science, building a model telegraph with me and teaching me about how everything worked (i.e., anything I asked about), like car engines, telephones, and so on.

So while I had to work hard in college and graduate school and sometimes did need to modify or fine tune my study skills, a lot of the skills that I developed as a young adult were skills I had been exposed to in one form or another while growing up. (For example, my mother would never spell or define words when I asked this of her. Instead, she always made me look them up in the dictionary.) Stephens, Hamenadi, and Destin (2014) would argue that my upper middle class background taught me the rules of the game, of how to be a successful college student. Bourdieu (Bourdieu & Passeron, 1970) would argue that growing up in an upper middle class family with a father who had a college degree in science provided me with the cultural capital that enabled me to succeed in college level science.

However, my awareness of the benefits and advantages I had growing up is why I am passionate about the topic of SRL and why I firmly believe that it is crucial that we expose students, especially first-generation science students, to its benefits. In my view, we owe it to first-generation college students to clue them in as to the rules of the game, to guide them in how to be successful college students. This is especially the case in STEM where they stakes are so high—where dropout and failure rates are notoriously high (Barr, Gonzalez, & Wanat, 2008; Chen, 2013; Malcolm & Feder, 2016) and where opportunities for economic improvement as a result of a STEM college degree abound (Langdon, McKittrick, Beede, Khan, & Doms, 2011).

References

Azevedo, R., & Cromley, J. G. (2004). Does training on self-regulated learning facilitate students' learning with hypermedia? *Journal of Educational Psychology, 96*(3), 523–535.

Barr, D. A., Gonzalez, M. E., & Wanat, S. F. (2008). The leaky pipeline: Factors associated with early decline in interest in premedical studies among underrepresented minority undergraduate students. *Academic Medicine, 83*(5), 503–511.

Bernacki, M., Vosicka, L., & Utz, J. (2016, April). *Can brief, web-delivered training help STEM undergraduates "learn to learn" and improve their achievement?* Paper presented at the Annual Conference of the American Educational Research Association, Washington, DC.

Bernacki, M. L., Vosicka, L., & Utz, J. (2017). *Web-delivered training to improve learning and achievement for underrepresented and first-generation STEM learners.* Paper presented at the Annual Conference of the American Educational Research Association, San Antonio, Texas.

Bourdieu, P., & Passeron, J. C. (1970). *Reproduction in education, society and culture* (R. Nice, Trans.). London: SAGE.

Brown, B. R., Mason, A., & Singh, C. (2016). Improving performance in quantum mechanics with explicit incentives to correct mistakes. *Physical Review Physics Education Research, 12*(1), 1–20.

Bunce, D. M., Komperda, R., Schroeder, M. J., Dillner, D. K., Lin, S., Teichert, M. A., & Hartman, J. R. (2017). Differential use of study approaches by students of different achievement levels. *Journal of Chemical Education, 94*(10), 1415–1424.

Cataldi, E. F., Bennett, C., & Chen, X. (2018). *First-generation students: College access, persistence, and postbachelor's outcomes* (NCES 2018-421). Washington, DC: National Center for Education Statistics, U.S. Department of Education.

Chen, X. (2013). *STEM Attrition: College students' paths into and out of STEM fields* (NCES 2014-001). Washington, DC: National Center for Education Statistics, Institute of Education Sciences, U.S. Department of Education.

Chen, P., Chaves, O., Ong, D. C., & Gunderson, B. (2017). Strategic resource use for learning: A self-administered intervention that guides self-reflection on effective resource use enhances academic performance. *Psychological Science, 28*(6), 774–785.

CUNY Office of Institutional Research and Assessment. (2014). *Student experience survey.* Brooklyn College, Brooklyn, NY.

Cook, E., Kennedy, E., & McGuire, S. Y. (2013). Effect of teaching megacognitive learning strategies on performance in General Chemistry courses. *Journal of Chemical Education, 90,* 961–967.

Davis, J. (2010). *The first-generation student experience.* Sterling, VA: Stylus.

Dennis, J. M., Phinney, J. S., & Chuateco, L. I. (2005). The role of motivation, parental support, and peer support in the academic success of ethnic minority first-generation college students. *Journal of College Student Development, 46*(3), 223–236.

Dibenedetto, M. K. (2010, May). *Do self-regulated processes such as study strategies and satisfaction predict grade point averages for first and second generation college students?* Paper presented at the annual meeting of the American Educational Research Association, Denver, Colorado.

DiBenedetto, M. K., & Bembenutty, H. (2013). Within the pipeline: Self-regulated learning, self-efficacy, and socialization among college students in science courses. *Learning and Individual Differences, 23,* 218–224.

Ege, S. N., Coppola, B. P., & Lawton, R. G. (1997). The University of Michigan undergraduate chemistry curriculum 1. Philosophy, curriculum, and the nature of change. *Journal of Chemical Education. 74*(1) 74–83.

Ertmer, P., & Newby, T. (1996). The expert learner: Strategic, self-regulated, and reflective. *Instructional Science, 24,* 1–24.

Grove, N., Hershberger, J. L., & Bretz, S. L. (2008). Impact of a spiral organic curriculum on student attrition and learning. *Chemical Education Research and Practice, 9,* 157–162.

Henderson, C., & Harper, K. A. (2009). Quiz Corrections: Improving learning by encouraging students to reflect on their mistakes. *The Physics Teacher, 47,* 581–586.

Hoffer, T. B., Dugoni, B. L., Sanderson, A. R., Sederstrom, S., Ghadialy, R., & Rocque, P. (2001). *Doctorate recipients from United States universities: Summary Report 2000. Survey of earned doctorates.* Chicago, IL: National Opinion Research Center.

Horowitz, G. (2011, November). *Encouraging academic help seeking: Improving student performance through action research.* Presented at the Center for Teaching, Brooklyn College, The City University of New York, Brooklyn, NY.

Horowitz, G., Rabin, L., & Brodale, D. (2013). Improving student performance in organic chemistry: Help seeking behaviors and prior chemistry aptitude. *Journal of the Scholarship of Teaching and Learning, 13(3),* 120–133.

Hudesman, J., Crosby, S., Ziehmke, N., Everson, H., Isaac, S., Flugman, B., & Zimmerman, B. (2014). Using formative assessment and self-regulated learning to help developmental mathematics students achieve: A multi-campus program. *Journal on Excellence in College Teaching, 25*(3&4), 107–130.

Jehangir, R. R. (2010). *Higher education and first-generation students: Cultivating community, voice, and place for the new majority.* New York, NY: Pellgrave Macmillan.

Karabenick, S. A. (2003). Seeking help in large college classes: A person-centered approach. *Contemporary Educational Psychology, 28*(1), 37–58.

Kim, Y., & Sax, L. J. (2009). Student-faculty interaction in research universities: Differences by student gender, race, social class, and first-generation status. *Research in Higher Education, 50*(5), 437–459.

Kniffin, K. M. (2007). Accessibility to the PhD and professoriate for first-generation college graduates: Review and implications for students, faculty, and campus policies. *American Academic, 3,* 49–79.

Langdon, D., McKittrick, G., Beede, D., Khan, B., & Doms, M. (2011). *STEM: Good jobs now and for the future* (ESA Issue Brief #03-11). Washington, DC: U.S. Department of Commerce.

Lovett, M. C. (2013). Making exams worth more than the grade. In M. Kaplan, N. Silver, D. LaVaque-Manty, & D. Meizlish (Eds.), *Using reflection and metacognition to improve student learning* (pp. 18–52). Sterling, VA: Stylus.

Malcolm, S., & Feder, M. (Eds.). (2016). *Barriers and opportunities for 2-year and 4-year STEM degrees: Systemic change to support students' diverse pathways.* Washington, DC: National Academies Press.

Malinak, S. M., Bayline, J., Logan, Brletic, P. A., Harris, M. F., Iuliucci, R. J., Leonard, M. S.,. . .Sunderland, D. P. (2014). The impacts of an "organic first" chemistry curriculum at a liberal arts college. *Journal of Chemical Education, 91,* 994–1000.

McGuire, S. Y., & McGuire, S. (2015). *Teach students how to learn.* Sterling, VA: Stylus.

Nandagopal, K., & Ericsson, K. A. (2012). An expert performance approach to the study of individual differences in self-regulated learning activities in upper-level college students. *Learning and Individual Differences, 22,* 597–609.

Newman, R. S. (2000). Social influences on the development of children's adaptive help seeking: The role of parents, teachers, and peers. *Developmental Review, 20,* 350–404.

Olszewski, P. (2016). Teaching millennials how to study under the 21st century sky. *Pyrex Journal of Educational Research and Reviews, 2*(1), 1–9.

Reay, D., David, M. E., & Ball, S. (Eds.). (2005). *Degrees of choice.* Sterling, VA: Stylus.

Rozell, T. G., Johnson, J., Sexten, A., & Rhodes, A. E. (2017). Correcting missed exam questions as a learning tool in a physiology course. *Journal of College Science Teaching, 46*(5), 56–63.

Sebesta, A. J., & Bray Speth, E. (2017). How should I study for the exam? Self-regulated learning strategies and achievement in introductory biology. *CBE Life Sciences Education, 16*(2). doi:10.1187/cbe.16-09-0269

Stephens, N. M., Hamedani, M. G., & Destin, M. (2014). Closing the social-class achievement gap: A difference-education intervention improves first generation students' academic performance and all students' college transition. *Psychological Science, 25*(4), 943–953.

Stephens, N. M., Markus, H. M., & Phillips, L. T. (2014). Social class culture cycles: How three gateway contexts shape selves and fuel inequality. *Annual Review of Psychology, 65,* 16.1–16.24.

Szu, E., Nandagopal, K., Shavelson, R., Lopez, E., Penn, J. H., Scharberg, M. A., & Hill, G. (2011). Understanding academic performance in Organic Chemistry. *Journal of Chemical Education, 88*(5), 1238–1242.

Thorne, R. (2013, October 1). 3 ways to help first-gen college students survive in STEM courses [Blog post]. Retrieved from https://sparkaction.org/content/3-ways-first-gen-survive-STEM-college

VanderStoep, S. W., Pintrich, P. R., & Fagerlin, A. (1996). Disciplinary differences in self-regulated learning in college students. *Contemporary Educational Psychology, 21*(4), 345–362.

Ward, L., Siegel, M., & Davenport, Z. (2012). *First-generation college students: Understanding and improving the experience from recruitment to commencement.* San Francisco, CA: Jossey Bass.

Weinstein, C. E., & Van Mater Stone, G. (1993). Broadening our conception of general education: The self-regulated learner. In N. A. Raisman (Ed.), *Directing general education outcomes. New directions for community colleges.* San Francisco, CA: Josey-Bass.

Zimmerman, B. (1990). Self-regulated learning and academic achievement: An overview. *Educational Psychologist, 25*(1), 3–17.

4

Why Incorporate Self-Regulated Learning Into Your Classroom

It's a "No Brainer"

Relax: There Is Very Little Time and Effort Required of You

So you may be thinking, I don't have the time, let alone the expertise, to train my students in how to study. Can't our tutoring center, advisement center, dean's office, or some other area within my university deal with issue? Not to mention where would I fit this sort of training into my already densely packed, content-focused course and curriculum?

Well. First of all, teaching students how to study more effectively (e.g., how to improve their self-regulation) is easy! It's a no brainer, as the title of this chapter implies. As the author of *Creating Self-Regulated Learners* (Nilson, 2013, p. 4) puts it, "Self-regulation has little to do with intelligence and just about anyone can develop it." You are actually an expert in how to study your own discipline, although you may not think about this consciously on a regular basis. And I'll bet that you would find explaining methods of

Teaching STEM to First Generation College Students, pages 31–33
Copyright © 2019 by Information Age Publishing
31

studying to students to be easier than explaining some of the most difficult, esoteric topics of your discipline.

Additionally, teaching students how to improve their study skills should *not* take up extra class time. Comments about how to study can be easily woven into your existing lectures or other presentation methods. If you look ahead to Chapter 5, you will see that in that chapter I detail many strategies you can utilize to more effectively guide your first-generation college students, many of which do not require extra class time, and some of which are just plain, good teaching practices.

You Are the Best Person for the Job

Okay. But you are probably still thinking, why should this be *my* job? Can't some office at my university run these trainings? And my answer is, yes, they could. *But*, research shows that stand-alone trainings (like "adjusting to college" type courses) are not very effective because it is difficult for students to figure out how to transfer the general ideas they learn about in these courses to the specific contexts in which they will need to apply them (Hattie & Marsh, 1996; Steiner, 2016; Wingate, 2006).

What do I mean by this? Well you could imagine that how one might study for a physics course might be different than how one might study for a biology course. Or, even that how one might study for one particular biology topic that is very fact based might be very different from how one might study for another biology topic that is very visual. So, study skills are very context dependent (S. Karabenick, personal communication, 2016).

Students will most effectively learn and adopt strong self-regulated learning skills and strategies when they are taught within the exact context in which they must utilize these skills. As Sebesta and Speth (2017), who study the correlation between self-regulation and biology performance, put it,

> While broad, nontargeted study skills seminars, workshops, or one-on-one tutoring are frequently offered by college and universities, they may not be the most effective...Instructional approaches that foster development of self-regulated learning habits need to be embedded within specific course contexts and appropriately reflect content and practices of the discipline... (p. 30)

So I would invite you to consider that, with a little bit of forethought and with very little time or effort, you can do a lot to help your first-generation college students navigate better through your courses, thereby improving the success rates of these students in your courses. See the next chapter for ideas about how to bring this about.

References

Hattie, J., & Marsh, H. W. (1996). The relationship between research and teaching: A meta-analysis. *Review of Educational Research, 66*(4), 507–542.

Nilson, L. B. (2013). *Creating self-regulated learners: Strategies to strengthen students' self-awareness and learning skills.* Sterling, VA: Stylus.

Sebesta, A. J., & Bray Speth, E. (2017). How should I study for the exam? Self-regulated learning strategies and achievement in introductory biology. *CBE Life Sciences Education, 16*(2). doi:10.1187/cbe.16-09-0269

Steiner, H. H. (2016). The strategy project: Promoting self-regulated learning through an authentic assignment. *International Journal of Teaching and Learning in Higher Education, 28*(2), 271–282.

Wingate, U. (2006). Doing away with "study skills." *Teaching in Higher Education, 11*(4), 457–469.

5

Advice for Lecturers

How to Incorporate Self-Regulated Learning Into Your STEM Course (or Just How to be a More Effective Instructor of First-Generation Students)

The Benefits of Working With First-Generation College Students

First-generation college students generally value education more than continuing generation students (some of whom take it for granted; Davis, 2010). They tend to be highly motivated and very hard working—the former because they know how education can vastly improve their economic situations and the latter because they often have to work very hard just to survive financially. So they can be a great joy to teach. But often, they struggle to succeed. The good news is that there is a lot you can do to help them!

Focus on the Positive

Without even being aware of it, you communicate unspoken messages to your students about your beliefs and expectations as to whether or not you

Teaching STEM to First Generation College Students, pages 35–52
Copyright © 2019 by Information Age Publishing
35

think they will be successful in your classes. Furthermore, a number of studies have demonstrated that your unspoken messages do matter—that your tone of voice, hand gestures, and overall attitude can influence the academic outcomes of your students (Flannery, 2015; Rosenthal & Jacobson, 1968). You therefore, are in a position in which you can convince students that STEM is something they can do if they put in the right amount and right type of effort. But you also have the power to send a different message to them—that STEM is something that requires a unique, innate talent.

So focusing on the positive, encouraging students, offering them guidance, making them aware of strategies that other students in your discipline have found helpful, are all ways in which you can indirectly improve the performance of your students. In particular, cluing in first-generation students about how to be successful in your discipline and sending them the message that you believe in them are ways that you can encourage their success.

Provide Guideposts Along the Way

Content

Introductory science courses often require students to assimilate large amounts of material in short amounts of time. You, as the content expert, however, can direct students as to how to organize the course material, as to how the pieces of the puzzle fit together (Ambrose, Bridges, DiPietro, Lovett, & Norman, 2010). You can guide them as to how to pace themselves and as to what to focus on (MgGuire & McGuire, 2015).

Specifically, when presenting material in class, you can let students know

- when a more difficult or more dense topic is about to come up,
- which topics are most crucial and important, and
- which topics are foundational and involve core, prerequisite material that will matter later on in the semester (as opposite to topics that are isolated, stand-alone material).

Keep in mind that first-generation students are going to be the least able to pace themselves. Letting them know what they should be working on at present and letting them know what is about to come next can go a long way towards helping them stay on track.

Skills

Similarly, you can let students know what skills they are expected to master (Ambrose et. al, 2010). For example, in my Organic Chemistry I course,

I tell my students that when it comes to the subject of stereochemistry, that there are a series of *four* skills they will need to master in order to perform successfully on exams and quizzes: (a) determining priority rules, (b) determining R/S configuration where position of group 4 varies, (c) determining relationships between isomers, and (d) converting from one type of 3D projection to another. I explicitly tell this information to my students because sometimes these skills are depicted separately in the homework problems, but sometimes they are combined in a way that is not always clear to students. Furthermore, each of these individual skills needs to be practiced separately and thoroughly mastered before problems can be solved that require the combination or synthesis of the skills.

As a subject expert, it may be obvious to you what skills are associated with a particular topic, but your first-generation students are the most likely to miss out on distinctions among the different skills. They will also be least likely to realize when they may have mastered only a portion of the skills required for your exams.

Homework

Assign specific homework problems to students. Don't expect students to be able to find problems on their own, nor expect them to be able to pick out appropriate problems from the textbook. You as the subject expert know best which topics are the important ones and which topics are most difficult for students. You need to choose the right problems for students, to focus in on what you feel is most important and to emphasize the areas you know will need particular attention and practice. You also need to scaffold homework assignments and assign students problems sequenced levels of difficulty so they can first master the basics before they then tackle more challenging problems. (For more about scaffolding and how it is used in educational contexts, read about the "zone of proximal development" or "Lev Vygotsky" in Wikipedia.)

Differentiate also for students between more "quiz" style questions that hone in on very specific topics as opposed to more broad questions (typically found on exams and finals) that integrate or bring together material from multiple chapters and require students to think *across* chapters in order to answer them.

If your textbook doesn't provide appropriate or sufficient homework problems for students, your savviest students will likely know how to find suitable practice problems on the internet. But it is likely that your first-generation students will not know how to do this effectively. In my experience when choosing problems on their own, often first-generation students cannot distinguish between problems whose topics they are or are not responsible for. They therefore waste lots of precious time struggling to solve problems on subject areas that have not yet been covered, that have been deliberately omitted or that been deemed unimportant by instructors. You can best help your first-generation students by selecting online resources for them that are a good match for your curriculum and your level of difficulty.

Provide Frequent Feedback

One way to let students know that they are or not keeping up properly in the course or that they have or have not effectively mastered a topic is to assess their learning (e.g., via a quiz or exam) and to provide information back to them about what they know, what they do not know, what they ought to be focusing on, how they should adjust their study strategies, and so on.

You might think that pop quizzes or even just frequent announced quizzes will keep the students on their toes and help them to keep up. But *feedback* and *prompt feedback* (e.g., providing a grade and/or comments and corrections on a quiz) is incredibly important (Ambrose et al., 2010). Research shows that weaker students tend to be overconfident and overestimate their exam performance while stronger students tend to underestimate their performance, a phenomenon known as the Kruger Dunning effect. (For a general description see Kruger & Dunning, 1999; for effect in college science courses see Bell & Volckmann, 2011; Hawker, Dysleski, & Rickey, 2016; Lindsey & Nagel, 2015; Pazicni & Bauer, 2014;). This means that the simple act of taking a quiz or exam may or may not be a wake-up call for the students who most need the wake-up call.

Therefore, it can be very helpful when assessing students to provide them with prompt feedback about what they are doing right and wrong. The assessment can take the form of a formal examination or quiz, but can be done effectively through informal methods such in-class polling where students vote on a multiple choice question (using clickers, a smart phone, holding up fingers or a colored card, or even orally). Alternatively, students can be asked to draw a diagram or structure (on paper, white board, or smartphone) and hold it up or submit it (see Horowitz, 2016). Depending on the nature of the course, students can also be asked to write a short summary or reflective statement which they can turn in for grading (see Davis, 2009, Chapter 32 for more on informal assessment assignments.)

Of course it is true that the more frequently you assess students, the more grading you have to do and the harder it will be to get feedback back to students quickly. A number of options are available to help with this problem.

- Posting an online answer key immediately after an exam or quiz while students are still hyped up and anxious about how they did can encourage students to investigate and discover for themselves what they know and don't know even if they don't have their graded papers in hand.
- Online homework and/or cognitive tutors that can customize feedback and instruction (Koedinger & Tanner, 2013) are a potential option as well, but it may unfortunately be the case that your first-generation students may be least likely to be able to afford these resources.
- Many students will be motivated to check their grades online even if you will not see them in person right away to return their graded work. Therefore, quickly entering grades into an online course management system is a convenient way to give students some information about how they are doing relative to how they perceive themselves to be doing.
- You can also use a course management system to insert and/or calculate a letter grade for each student to give them an approximate idea of how they are doing in your course at different points in time. (This can be dicey because students may try hold you to the approximate grades you enter and you yourself may not want to commit early on to specific grade cutoffs. However, I have found that when I don't assign an approximate letter grade, my least clued in students are most likely to overinflate how they are doing and not realize when they are in danger of failing. So entering approximate letter grades is a possible option in this situation. Alternatively, it is possible to just put a warning type letter grade in only for students who are in serious trouble.)

Provide Opportunities for Self-Reflection

A number of STEM faculty have begun to make use of something called *exam wrappers* (see Chen, Chavez, Ong, & Gunderson, 2017; and Chew, Chen, Ricken, Turpin, & Sheppard, 2016 for examples of successful STEM interventions). Exam wrappers, designed by Dr. Marsha Lovett of Carnegie Mellon University (2013), are a written exercise or activity intended to be filled out by students following a quiz or exam. (Often students are asked to fill out exam wrappers at the time their graded exams are returned to them.) Exam wrappers are designed to encourage students to review their exam or quiz paper, their errors, their overall performance, and to reflect on their studying. Students are asked to describe in writing whether or not they feel satisfied with how they prepared for a particular exam or quiz and to write about how they might prepare differently in the future for the next exam or quiz.

In my own teaching, I have used exam wrappers both after students receive exams or quizzes back (in a posttest like manner), but also as a *pretest* exercise where immediately preceding a quiz, students are asked to fill out a short survey where they assess their knowledge or their confidence in their knowledge of the course material. The purpose of the pretest exam wrapper is to shake students out of their complacency, as often it is the case that weaker students overestimate their knowledge of the course material. (This phenomenon is known as the Dunning-Kruger effect (Kruger & Dunnger, 1999) and is frequently observed in introductory STEM courses (Bell & Volckmann, 2011; Lindsey & Nagel, 2015; Pazicni & Bauer, 2014). So I want to try to help make my students more conscious and aware of the fact that they are doing this by administering pretest exam wrappers to them.

Examples of exam wrappers for STEM classrooms can be found at the Carnegie Mellon website (https://www.cmu.edu/teaching/designteach/teach/examwrappers/) and in Chapter 2 of *Using Reflection and Metacognition to Improve Student Learning* (Kaplan, Silver, Lavaque-Manty, & Meizlish, 2013).

Exam wrappers can benefit all students and encourage all students to be more self-reflective about their study habits, but exam wrappers can be especially beneficial to first-generation college students who need the most guidance and feedback about their study habits.

Write Effective Exams and Quizzes . . .
Especially for Newer Faculty

As I have described above, ideally the purpose of assessment should be to provide feedback to both yourself and your students as to what extent they have mastered the content and skills that you wished them to master. It is important therefore that your exams and quizzes truly test or assess what you want them to.

Provide Clear Instructions

For example, if your exam requires that students fill out a scantron or some other form or format, make sure the instructions are clear as to what method students use to enter their answers. I once proctored a scantron exam in which one student circled his or her answers on the question sheet and did not fill out the scantron even though pencils were provided. Perhaps this student was unfamiliar with scantrons or did not understand the exam instructions. But as a result, instead of his or her knowledge of chemistry being assessed, his or her familiarity with an assessment system was measured. This is not to say that students should not be held responsible for knowing the academic system in which we function, but if your goal in assessing students is to learn about their mastery of a STEM subject, make sure your assessment instrument does this and does it well.

The following are some specific, practical suggestions as to how to write an exam so that it assesses what you really want it to assess.

- Make sure all exam instructions are written clearly and correctly.
- Make sure all exam questions are worded carefully, clearly, and in grammatically correct English. Make sure they have no errors or typos in them.
- Try to display your questions in the clearest possible layout (e.g., think about font, line spacing, paragraph spacing, etc.). Leave enough empty space on each page to make sure your layout is clear even if this means "wasting" paper.
- Try not to have a single question run over more than one page.

Do a Sample Run Yourself

One way to systematically check your exam for layout, presentation, clarity, and errors is to go through your exam yourself and thoroughly complete each answer as if you were a student attempting to get full credit on each question. Not only will you notice errors or ambiguities this way, but you can get a good measure of how long it will take students to complete your exam. Figure that it will take students 4 to 5 times longer than you to complete your exam and time yourself to see how long it takes you to do the exam. (*Note:* Answering the questions of your own exam while timing yourself kills two birds with one stone because it is also a convenient way of creating an answer key for your exam.)

If you determine that your exam is too long, trim or edit the exam. "Why?" you may ask. An exam that is too long doesn't do a good job of assessing what students know. Rather it gauges who is most familiar with competitive, timed exams, and it only gauges learning for those questions that students have time to complete.

Discuss Exam Format in Advance

Students may do homework problems for your class, but these homework problems may or may not follow the format of your exams. Let students know in advance what format or formats you will utilize (e.g., essays, short answer, matching, multiple choice, calculations). If the format you use is unusual and not something they will find in their textbook, provide them with examples of the format(s) you will use. Let students know if they will be required to show their work, if they will need to bring a calculator (what type of calculators will be permitted), or if there is a particular format in which they must show their answers.

You may be thinking that you shouldn't need to coddle students by providing them with this type of information. But I would argue that if you write a question that requires students to understand an unfamiliar question format, that your question is wasting both time and space on your exam, as it will not end up assessing the content you wished to assess. Instead, you will end up assessing students' savviness and familiarity with different types of exam formats.

For example, in my discipline (organic chemistry), students are expected to answer a variety of questions in which they must write down chemical structures or chemical names (mechanism questions, synthesis questions, and road map questions). Usually students are expected to figure out on their own how these three types of questions differ from one another. For example, in road-map questions, students must fill in blank spots with the structures of *all* missing compounds (starting materials, reagents, intermediates, and products). Yet, in a synthesis question, students are usually *instructed* to fill in *reagents only*, yet are also *expected* to fill in stable *intermediates*. And with mechanism questions, there is an unwritten rule that students *must not* write down any reagents at all. Rather, students must show a series of unstable intermediates and show the motion of electrons with arrows in order to receive credit for this type of question.

Yes, students must learn to differentiate between these types of questions. But, I think it is reasonable to give them a heads up that these different types of questions exist and that they each require practice of different skills. I also think it is reasonable to explain what mechanism or road maps questions are. This hopefully makes sure that students understand what is being asked so that when grades are obtained for these students, the grades measure the skills or content that one wished to assess.

Allocation of Points and Time

It is also important to carefully allocate points on exams. It is easiest to grade an exam when all questions are worth an equal amount (e.g., 10 questions each worth 10 points). But, this does not accurately and fairly provide feedback to students nor you as to where the students are at in terms of mastery of the course material. A difficult or time consuming question ought to be given more weight than an easy one. And a topic that you feel is important ought to be weighted more heavily than a topic you feel is trivial.

As I stated above, it is a good idea to make sure your exam is doable in the allotted time; otherwise you are just testing students regarding their experience with test taking and fluency in English rather than their STEM mastery. Because of this, I always include the point allotment of each question next to each exam question I write. Recently however, I have also gotten in the habit of including an approximate time allotment next to each question (indicating how long students should spend on each question). I do this to help students train themselves to pace themselves on timed exams, to train themselves to not devote too much time to questions that are worth very little and to set aside enough time for questions that are worth a lot.

Question Difficulty

I also try to write questions of varied difficulty so that my exams can best resolve students one from the next in terms of skills and knowledge. Some faculty write exams with all easy or all very difficult questions. But I feel that both of these types of exams are ineffective at resolving individual students one from the next. Very easy or very difficult exams also do not provide useful feedback to students regarding their strengths and the areas in which they need to improve.

Disseminate Information Broadly and Repeatedly

One of the areas in which first-generation students may be at the greatest disadvantage is their ability to understand and assimilate the rules associated with the course. Often these are the rules you lay out in your syllabus and/or discuss orally on the first day of class.

But because these types of rules are such an important part of a course, I recommend that whenever you disseminate crucial information to students (e.g., the day, time, and place of an exam or what material is covered on the next quiz) that you disseminate this information in multiple ways and to multiple destinations. Even if your in-class attendance is high, it is inevitable that students become ill, have job and family emergencies, or just have to make a quick run to the restroom while you are announcing crucial course information. And even students who are physically present will not catch or absorb everything you say the first time around. Students need time to assimilate things you announce.

Furthermore, keep in mind that often students are still adding and dropping courses during the first week of the semester. They may not enroll in your class right away, might miss the first crucial class, and may not be able to access course management software until they are on the roster.

I therefore advise that you try to put all important information in writing in your syllabus and that you don't tamper with or change anything in your syllabus except under extreme circumstances. Also, do not just post your syllabus online. Hand out hard copies of your syllabus on the first day of class so that students can follow along and jot down notes while you go over the syllabus during the first class period. (You can also post an electronic copy in addition to handing out hard copies.)

Furthermore, I advise when posting course related information electronically that you use your webpages (or departmental webpages) to do this, rather than course management software. This will enable students who may not be registered yet or who are having troubling logging in to the course management software system to be able to view the information. (I only use course management software for posting grades or copyrighted materials.)

In addition to the syllabus, there will often be information that you need to announce more spontaneously in class (e.g., a change in exam schedule, an extra review session, etc.) An oral announcement at the beginning of class may be missed by latecomers, but may also not get absorbed by everyone present. Even a PowerPoint slide at the beginning of class may be missed or not taken in by all students. I therefore advise posting announcements on a corner or side blackboard so that they can remain up throughout the class. If you always post announcements on the same corner blackboard each day, students will quickly learn to look there to find out important information, changes, and reminders. Announcements can also be disseminated in additional ways by posting them on your website and/ or by emailing the class listserv. (Usually this can be easily done with course management software.) Using your website to post (rather than course management software) will enable students to quickly check to see if there is anything new posted without needing to login every time they want to check/look for an update.

If at all possible, I would also include information about resources and sources of help in both your syllabus and your website. As my co-authors and I discuss elsewhere (Horowitz, Rabin, & Brodale, 2013), more savvy students are more likely to take advantage of resources and help. First-generation students may be the least familiar with college resources and feel the least entitled to take advantage of resources (feel least empowered to partake of college resources).

Don't Be Afraid to Follow-Up or Pursue

Diagnostic Conferences

After I give back a (graded) exam, I deliberately hold extra office hours (with fixed appointment times that students can sign up for) to give students an opportunity to meet with me one-on-one to discuss how they are doing the course. I find that these meetings are very useful because it is often only when students have cold, hard evidence in front of them, that they are open to hearing about changing their study habits.

When I have these meetings with students, often they want to know why they got a particular question wrong; sometimes they also try to haggle for points. And occasionally they do find grading errors. So I do try to first have students review the answer key online before we meet so as to not waste too much time haggling over points. And I also have an online form that they use to submit a regrade request if they feel an error was made in the grading of their exam (Horowitz, 2016).

I utilize the online answer key and regrade request form to get the mundane issues out of the way so that when I meet with students, I can direct the conversation to deeper, more fruitful issues. I treat this meeting almost like a consultation or intake visit one would have with a doctor in which I, playing the role of the doctor, collect information or evidence from the student, then assess, and then offer suggestions in the form of different types of "treatments" or "prescriptions" regarding study habits or change in study habits and study strategies.

I look for information or evidence in a few different ways when I meet with each student.

1. Before the student walks in, I pull up a record of the student's scores (all quizzes and exams) in the course management software.
2. If I see that the student did okay, but not terrific, on the exam, I start by looking over the student's exam and look for trends. What type of questions are they frequently getting right and getting wrong? Are there topics in which they are stronger, weaker, or need to review? Based on what I observe and what I hear from the student, I do one or more of the following:

- Sometimes I assign the student specific problem sets off of my website (Horowitz, 2016) that target specific topics—to help the student shore up his/her skills on particular topics.
- If the student seems to have more major issues where they are struggling to understand core fundamentals, I often suggest they see a tutor. I then hand them a sheet of information about tutoring. The sheet has our learning center's schedule of tutor hours (free) on it and it has information about private (not free) tutors that I know and recommend.

3. If I see from the student exam scores that this student is in real trouble (failing), then I start the conversation differently by asking the student what is going on with them, what happened, where they think they went wrong?
   - Sometimes students describe very serious events going on in their lives that got in their way and then we talk about if it is possible for them to recover and catch up at that point in the semester. Sometimes I go on to recommend tutors and hand them my tutor info sheet, but other times we discuss the pros and cons of them withdrawing.
   - And because many of my first-generation students have major, serious things going on in their lives, I often try to encourage students to get support for what they are going through. I often recommend they find a confidential ear by going to see someone in our personal counseling center. But because I know that there can be a stigma associated with seeing a counselor, I try to make this suggestion gently and emphasize that this resource is something free, that their tuition money already pays for, and that this resource is a nonjudgmental, confidential ear that they will not have to face again (unlike family, friends, instructors). Additionally, since it's free, it costs nothing to try it once. (If they don't like it, then don't have to do back.)

4. But most often when I meet with a student who is failing, it is not because of all the heavy stuff going on in their life (which can get in the way of them putting in time to the course). Sometimes I discover that it is simply because of test anxiety (see discussion regarding pretest anxiety below). But most of the time, I discover that it is because *they are not studying correctly.* And sometimes these two phenomena are coupled (anxiety and improper studying) because being unprepared for an exam often leads to test anxiety.

So when asking students who are failing or are close to failing what happened to them that they were not successful, I usually follow up with a question about their study habits. Almost all failing and borderline students tell me that they are studying and studying hard (typically 10 to 15 hours per week for my course alone). But I then follow up by asking *how* they are studying. And most frequently I find that students who report that they are working very hard, yet are doing very poorly, *are simply studying in the wrong way!*

Retooling of Study Habits and Behaviors

In the case of organic chemistry, most students get bogged down trying to read, understand, and digest our textbook. This leaves them little or no time for practicing problem solving, which is crucial for being able to perform on exams. I therefore start by telling students that they need to spend 80% of their time doing problem solving (e.g., 8–12 hours per week) and that they must figure out how to review the content of the course (e.g., 2–3 hours per week) so that they can move as quickly as possible to problem solving.

However, I find that a number of specific issues regarding study habits come up over and over again with students. Here is how I address each of them:

How to digest and understand the course material so you can be ready to solve problems:

1. Spend 5 minutes before class skimming the textbook so you have some idea of what will be covered in class.
2. Review your lecture notes after class. Fill in information you didn't have time to write down during class. Figure out which things you are still confused about and note this in the margin of your paper so you can quickly find it later.
3. For help with short, quick things:
   a. Go to office hours (lecturer, recitation instructor, or lab instructor) for help. (Here I give the student a copy of all the recitation instructors' hours if this is not available in the lecture syllabus.)
   b. During a waiting time in lab, ask the lab instructor for help.
   c. Ask classmates around you for help (in lecture, recitation, or lab).
   d. Go to the learning center.
4. For help with big topics:
   a. Find an online video to explain the topic (e.g., YouTube, freelanceteacher.com, khanacademy.org)
   b. Get a tutor privately or through the learning center.

What to do when you get stuck while problem solving:

1. Recognize that *everyone* gets stuck. This is nothing to be ashamed of. The people who seem to know what they are doing got confused by the same stuff as you. They just knew how to get unstuck by seeking help effectively.

2. Try to work/study in a group and try to find a group that is diverse in its knowledge (different people know different stuff) so people can help, but also challenge one another.

3. Try to do some problem solving every day. But just plan to do a little bit. And plan a help resource you can access each day so that you can get help and get unstuck. Review the office hours and learning center hours and plan a checkpoint for yourself each day so that you can get your questions answered on a daily basis so that you can move forward in your problem solving. *Never* plan to spend one whole day per week as your only time for problem solving. You could get stuck one hour in and not be able to move forward or even worse waste the whole day spinning your wheels/banging your head against the wall trying to get through one problem. If you get stuck and don't have a source of help available, don't waste time that you cannot afford to waste. Find something else to do and seek help as quickly as you can. You need to be comfortable admitting that you need help and asking for help. This is not a sign of weakness. It is a sign of being a successful student.

Anxiety and Panic Attacks:

If a student reports that they study yet panic and blank out during exams, it is often because they simply have not practiced properly or have not practiced enough.

1. I tell students to make sure they have practiced diverse problems, problems of different levels of difficulty (not just the "in the chapter problems," but also the hardest problems) and to make sure they practice problems that synthesize across topics.

2. Make sure they practice enough—practice till they are bored to death and then they will know they are ready.

If a student is overpracticed and still panics or blanks out during exams, then this student does have serious, diagnosable test anxiety. I would encourage this type of student to seek counseling or even consider consulting a physician.

Students Who Don't Show Up

In spite of my best efforts to encourage students to see me, I find that often it is the students who are most urgently in need of help who do not come in to see me. If this is the case for you as well, you may likely feel frustrated just like I do and may be wondering what is the use of a diagnostic conference if those who really need it don't show up?

This is a good question. I want to answer it but first state that when students don't do something (don't show up, don't put in effort, etc.), I find that 99% of time (especially with first-generation college students) it is not because of apathy, laziness, or a lack of motivation. It is an *active* avoidance. Students who are failing or near failing sometimes deliberately avoid any and all evidence or information that will confirm for them that they are in trouble. Why? They feel awful and feel ashamed. They do not want to confront the fact that they are failing. But they care very much about their situations, otherwise they would not need to avoid.

How do they avoid? Well, they want to avoid any evidence that they are in trouble. So they avoid coming to class to avoid hearing the lecture and being reminded of how lost they are. They avoid picking up quizzes or exams because they don't want to see all the red marks and minus signs on them. They come late to quizzes and exams so they can rationalize that their lateness is the reason they didn't do better. They avoid office hours and they avoid your post-exam diagnostics. They may even avoid responding to your email request that they come see you.

What can you do about avoiders?

1. Help students to understand (talk about this in lecture when discussing the exam performance of the class) that if they are doing poorly that no one is to blame. Rather most likely they need to "retool" their behaviors and that by having a sit down with you, you will come up with a game plan (diagnosis and treatment plan) for them.
2. Remind students that likely everyone feels lost in class sometimes even if they don't "look it."
3. Remind students that all students find your course difficult, but that you have experience with strategies that can help them.
4. If necessary, explain to students that it is not them; they are not smarter or stupider than others but that some may have had a better preparation from high school.

How to get avoiders to come see you?

First accept the fact that avoiders will not show up simply because you announce in class that you are available to meet with them.

1. An explicit, personal invitation is one way to start. Students may not realize that you cannot "require" them to come see you. Writing on an exam "Please see me!" may get some students to come see you. An email along the same lines cannot hurt. And I have even been known to track down students in lab (I look up their schedule online or contact the lab instructors), bring my appointment book with me to the lab, call the student out into the hall and just say, "I'm worried about you. Can we schedule an appointment to talk?" (None of these methods is foolproof, but all can be helpful.)

2. Use an online scheduler so that students can anonymously sign up for an appointment and can look over your schedule and availability in private.

3. Figure out something the student needs from you and use this as an opportunity to have the diagnostic conversation *you* want to have. Students who are in trouble may contact you, usually via email, to ask for something (e.g., to not count an exam or assignment, to get a makeup exam, to get an extra credit assignment, to drop your course after the deadline, to pass/fail the course, etc). Even if you know *in advance* that your answer to their question is "No," it can work well to respond to them by saying, "This is a topic we need to talk about in person." If and when they do show up, you can then direct the conversation towards the more fruitful, diagnostic conversation you want to have. The best way to ensure a productive conversation is to start off by letting the student know that you are generally concerned about them and invested in their future success (while also being truthful that the miracle or life preserver that they are hoping for is not going to happen).

4. I have even had students finish my course and get a D or F and then come back to beg for mercy. I have been successful with many of these students (although I do not show mercy) to reframe the conversation around what happened to them and why and what they can do differently to avoid this happening again. Usually for these students (if you are able to access and view their transcripts), your grade of D or F is not their first time struggling to pass a course (especially if they are a chronic avoider). But your conversation with them may very well be the first time they have sat down with a person with science expertise (e.g., a professor) to talk about what is happening to them, why, and what they can do about it.

Going the Extra Mile for Students

In the book *Teaching and Learning STEM*, Felder and Brent (2016) describe a number of other strategies for professors who want to go the extra mile to reach students, whether these be students who are in trouble or just all students in general. They recommend trying to meet with each student during the first week of the semester even if it is just for 5 minutes. They also recommend requiring students to come to your office to pick up their midterms by hand so that they are forced to discuss with you in person how they have done and what they can do to be more effective and more successful the next time around.

References

Ambrose, S. A., Bridges, M. W., DiPietro, M., Lovett, M. C., & Norman, M. K. (2010). *How learning works: Seven research-based principles for smart teaching.* San Francisco, CA: Josey-Bass.

Bell, P., & Volckmann, D. (2011). Knowledge surveys in general chemistry: Confidence, overconfidence, and performance. *Journal of Chemical Education, 88*, 1469–1476.

Chen, P., Chaves, O., Ong, D. C., & Gunderson, B. (2017). Strategic resource use for learning: A self-administered intervention that guides self-reflection on effective resource use enhances academic performance. *Psychological Science, 28*(6), 774–785.

Chew, K. J., Chen, H. L., Rieken, B., Turpin, A., & Sheppard, S. (2016, June). *Improving students' learning in statics skills: Using homework and exam wrappers to strengthen self-regulated learning.* Paper presented at the American Society for Engineering Education (ASEE) 123rd Annual Conference, New Orleans, LA.

Davis, B. G. (2009). *Tools for teaching.* San Francisco, CA: Jossey-Bass.

Davis, J. (2010). *The first-generation student experience.* Sterling, VA: Stylus.

Felder, R. A., & Brent, R. (2016). *Teaching and learning STEM.* San Francisco, CA: Jossey-Bass.

Flannery, M. E. (2015, September 9). When implicit bias shapes teacher expectations. *neaToday.* Retrieved from http://neatoday.org/2015/09/09/when-implicit-bias-shapes-teacher-expectations/

Hawker, M. J., Dysleski, L., & Rickey, D. (2016). Investigating general chemistry students' metacognitive monitoring of their exam performance by measuring postdiction accuracies over time. *Journal of Chemical Education, 93*, 832–840.

Horowitz, G. (2016). Comment on "Beyond clickers, next generation student response systems for Organic Chemistry." *Journal of Chemical Education, 93*(11), 1829.

Horowitz, G. (2016). Organic Chemistry I–II [Homepage]. Retrieved from http://userhome.brooklyn.cuny.edu/ghorowitz/

Horowitz, G., Rabin, L., & Brodale, D. (2013). Improving student performance in organic chemistry: Help seeking behaviors and prior chemistry aptitude. *Journal of the Scholarship of Teaching and Learning, 13*(3), 120–133.

Kaplan, M., Silver, N., LaVaque-Manty, D., & Meizlish, D. (Eds.). (2013). *Using reflection and metacognition to improve student learning.* Sterling, VA: Stylus.

Koedinger, K. R., & Michael, T. (2013, July 9). 7 things you should know about intelligent tutoring systems. *Educause Learning Initiative.* Retrieved from https://library.educause.edu/resources/2013/7/7-things-you-should-know-about-intelligent-tutoring-systems

Kruger, J., & Dunning, D. (1999). Unskilled and unaware of it: How difficulties in recognizing one's own incompetence lead to inflated self-assessments. *Journal of Personality and Social Psychology, 77*(6), 1121–1134.

Lindsey, B. A., & Nagel, M. L. (2015). Do students know what they know? Exploring the accuracy of students' self-assessments. *Physical Review Special Topics—Physics Education Research, 11*(2). https://doi.org/10.1103/PhysRevSTPER.11.020103

Lovett, M. C. (2013). Make exams worth more than the grade. In M. Kaplan, N. Silver, D. LaVaque-Manty, & D. Meizlish (Eds.), *Using reflection and metacognition to improve student learning* (pp. 18–52). Sterling, VA: Stylus.

McGuire, S. Y., & McGuire, S. (2015). *Teach students how to learn: Strategies you can incorporate into any course to improve student metacognition, study skills, and motivation.* Sterling, VA: Stylus.

Pazicni, S., & Bauer, C. F. (2014). Characterizing illusions of competence in introductory chemistry students. *Chemistry Education Research and Practice, 15*, 24–34.

Rosenthal, R., & Jacobson, L. (1968). *Pygmalion in the classroom: Teacher expectation and pupils' intellectual development.* New York, NY: Holt, Rinehart, & Winston.

6

Advice for Graduate Student Instructors

This Chapter Is for You

If you are a graduate student, working towards your master's or doctoral degree in a STEM (science, technology, engineering, mathematics) field, this chapter was written for you. You may be someone who looks forward to your teaching assignments—perhaps you are planning for a career in teaching. On the other hand, you may be someone who approaches teaching with fear and trepidation, wishing that teaching was not something required of you.

Either way, it is highly likely that your life as a graduate student is filled with a good deal of stress and that you often feel burdened with many responsibilities. You are likely juggling a combination of your own coursework, your research responsibilities, and your teaching responsibilities.

This chapter is here to help. This chapter will help you better understand the students you work with or will be working with. Understanding your students will make both teaching easier and preparing for teaching easier. And, it will make you a better teacher too!

Teaching STEM to First Generation College Students, pages 53–66
Copyright © 2019 by Information Age Publishing
All rights of reproduction in any form reserved.

The American Educational System

Many STEM graduate students who study in the United States come here from other countries to conduct their graduate work at U.S. universities (National Foundation for Educational Policy, 2017). Often, international graduate students come to the United States having already completed their undergraduate studies and even their master's level coursework in other countries.

For international graduate students, the initial stages of teaching in a university in the United States can be difficult and can require significant adjustment, in part because our primary and secondary educational systems are quite different from those found in many other countries. Even our terminology can be confusing. For example, we refer to primary schools as elementary schools, and to secondary schools as high schools. We typically refer to our combined primary plus secondary educational systems (elementary plus high school systems) as the K through 12 educational system or the *K–12* system for short.

The U.S. K–12 system educates over 90% of the K–12 age population. This is in contrast with education systems in other countries where often less than half of the K–12 age population finishes high school. For example, in India only 35% of young people finish high school, while in China only 52% do (UNESCO, n.d.).

The percent of college aged students who enroll in higher education (tertiary level education) is also much greater in the United States as compared to some other countries. For example according to 2015 data, 47% of persons aged 25–34 in the United States had earned an associate degrees or higher, while in China and Germany only 18% and 30%, respectively, had (OECD, 2016).

One might assume therefore that it is simply easier or less competitive to get into a university[1] in the United States as compared to other countries. However, what is in fact true is that it is very difficult and competitive to get into *elite* universities in the United States and unfortunately students' socioeconomic backgrounds play a significant role in terms of who gets into the most competitive, elite, and prestigious institutions in the United States (Carnevale & Strohl, 2010).

There is also another aspect of the U.S. educational system that is very different from that of other countries. The United States does not have a national K–12 educational system, with nationwide examinations at different stages along the way. There are rather many, many different educational systems in place across and within different states, across different districts within each state, and across different cities (Carnevale & Strohl, 2010).

Thus, while the vast majority of students in the United States receive a public (and free) K–12 education, the quality of this education can differ vastly from one local to another. In many locations, the amount of money spent on each pupil is heavily influenced by the taxes collected by the local government (e.g., the city or town; Bowen, Kurzweil, & Tobin, 2005), such that richer areas have vastly superior educational systems as compared to poorer areas (Augenblick, Meyers, & Anderson, 1997). In richer areas, schools have smaller class sizes, better quality teachers, more advanced science and math courses, more college advisors, and better classroom supplies and resources (Carnevale & Strohl, 2010).

How Does This Impact You?

You need to be prepared for the fact that the academic preparation and educational backgrounds of your students will differ one from the next based on their economic backgrounds and circumstances. And while your students may all look and dress like they come from similar economic circumstances (likely most students will try to look and dress in the right way so as to fit in with the culture of your institution; Landers, 2018), some students will be much better prepared academically than others. This diversity in background may give you the false impression that some of your students are much smarter or much more motivated than others. But *be careful* about making quick assumptions about your students before getting to know them better. I will discuss this topic further in the sections that follow below.

Your Teaching Assignment

It is likely that you will be teaching or serving in some other type of a supporting role, in an *introductory* course in your department.

What Does This Mean?

You may end up as a laboratory instructor, a recitation instructor (an instructor for a problem session), a grader, or in some other type of supporting role in what may turn out to be a very large lecture or laboratory course that has one full-time faculty member (a professor) at the helm, where you will be one of a group of graduate student instructors required to assist with the workload of the course. The professor will most likely rely on you to do much of the nitty gritty, busywork of the course. However, you as a graduate student instructor, will also likely have the most direct contact with the enrolled students.

Student Motivation

If you are teaching or working with a large introductory course, you may soon begin to notice that the motivation of your students may be somewhat different from your own. Most students who take large introductory courses take these courses to fulfill prerequisites so that they can move forward to study the subject area that they are interested in pursuing (which may or may not match up with the subject matter of the prerequisite course). These students, therefore, may not be very interested in the subject matter of the introductory course. This may be hard for you to grapple with as this subject (roughly speaking) is something you have chosen to pursue for your graduate degree.

However, keep in mind that what can help students stay interested and motivated in the subject, especially first-generation college students, is if you can demonstrate for them the relevance of the subject to their daily lives and/or to their career interests (Gabriel, 2018; Jackson, Galvez, Landa, Buonora, & Thoman, 2016; Priniski, Hecht, & Harackiewicz, 2018). Also, remember that in introductory courses, students are often bombarded with vast amounts of material all at once. The breadth of an introductory course can make it difficult for students to see the inner depths of the subject that you know to be interesting and fascinating. See if you can find ways to sometimes share glimpses with students of the more interesting parts of the subject to help them see that there is more to the material that just the broad bombardment they get exposed to in the introductory course.

Student Behavior

You will also find a mixture of behaviors among your students. Some students will seem to exhibit all the "right" behaviors: They will come to class on time, they will appear to be well prepared for class, they will have done the homework, they will turn in assignments on time, they will participate in class, they will attend your office hours. And the students who seem to do all the right things, will also often be the students who do well.

But what about the other students, the students who do not exhibit these ideal behaviors? What about the students who have spotty attendance, who come late, who turn in assignments late, who do not participate in class, and/or who do not attend office hours? Are they just lazy or unmotivated students?

Often, these negative conclusions are what new (and even some experienced) instructors tend to assume when they see behaviors that don't look like "good" student behaviors, behaviors that most STEM instructors would agree are not ones conductive towards success in STEM.

Making Assumptions Based on Behavior

It is quite natural, when we observe students acting in a way that it is not optimal for their own success, to assume that such students are lazy or don't care (are not motivated). In fact, generally speaking, we have a natural tendency to jump in (often too quickly) and attribute causes to others' behaviors without stopping to step back, think like scientists, and check, gather the facts, and find out what is really motivating a person's behavior (McGarty, Yzerbyt, & Spears, 2002).

To put it another way, we tend to quickly jump to conclusions when we observe behavior. We do this, because we are wired this way. To some extent, we have to think "on our feet" to survive (McGarty et al., 2002). We have to make snap decisions in our daily lives, whether it is a split second decision to apply the brakes of a car, or a quick judgment about how to respond verbally to an inappropriate comment of a student.

But when it comes to first-generation students, our judgments about their behaviors can often be faulty and can put these students at an educational disadvantage.

Let me illustrate with an example.

Back in January of 2018, I scheduled a research meeting with an undergraduate student and a colleague, who is a statistician. (The undergraduate was going to be doing a number of analyses for me using SPSS.) The purpose of the meeting was to discuss with the statistician which approaches we should take for our analyses. We had scheduled the meeting for 10:00 a.m. and scheduled to meet at the office of the statistician. At about 9:50 a.m., my student texted me to indicate that she would be a few minutes late, that she was having traveling difficulties. (I think she stated that she had missed her bus.) A good deal of time went by and the student did not show up. The student did not contact me again and I was not able to reach her by phone either. Eventually, my colleague and I started the meeting without the student. We spoke and worked on planning for a while. Eventually, the student showed up at approximately 10:35 a.m.

Based on the information you have so far about this student and the situation, what might you conclude about this student? Was this a serious, conscientious student? Or was this student a flake? Yes, she texted that she was running late, but her text implied she would arrive much earlier than she did. And she did not communicate again when her arrival time changed.

What do you think? How do you think I should have responded to this student when she arrived?

Well, let me share some additional information.

When my student arrived, we continued the meeting. Then after the meeting concluded, I asked her what happened and she said, "My phone died." She elaborated that she had indeed arrived at the correct building of my colleague shortly after 10:00 a.m. like her text message had said she would, but because the room number of my colleague (whom she did not know and had never met) was only contained in a text message on her phone, she had no way of figuring out where we were meeting.

So what did my student do? She searched for me in the only way she could think of. She walked through the hallway of *every floor of the building* for over half an hour until she heard my voice through the open doorway of my colleague's office. (There are approximately 100 rooms in that building.)

I realized a few things then and there. Firstly, that this student was not a flake or irresponsible person. She was a determined and resilient person who had learned how to be resourceful and not give up when faced with a challenge. I also realized that my student had an old cell phone battery, that didn't hold much charge anymore, and that a more well off person might have just purchased a new battery to avoid this problem altogether. And, that when my student had arrived on campus, she didn't have the option of just purchasing a new, extra phone charger at the local drug store like a more economically well-off person (like me) might have done. (I did have prior knowledge about the economic background of this student.)

Now you may be thinking there are other things this student could have done to prevent or anticipate this situation. But, the point of the story is that often when we observe behavior without knowing the cause we tend to blame the person in question (Ross, 1977), and when we know the cause we tend to be much more forgiving.

First-Generation College Students and Their Behaviors

You may be wondering what a story about a student whose cell phone battery died has to do with teaching first-generation college students. Or how we judge or don't judge people is relevant to your teaching. Let me try to explain.

First-generation college students (students whose parents did not attend or finish college) often lack savvy or sophistication regarding how to be successful in college level courses, particularly in STEM courses which require a specialized set of study skills (Davis, 2010; Jehangir, 2010; Thorne, 2013). Therefore, when you observe a student who does not participate in class, or does not seek out help during your office hours, or does not take you up on your offer to "email me with your questions," it is important not to leap to conclusions or judgments as to why this student does or does not take action. Often times, first-generation college students do not feel a sense of confidence in themselves (Nichols & Islas, 2016). They do not know that they belong in college and do not necessarily feel that they deserve the time and resources that are available to them (Rheinschmidt & Mendoza-Denton, 2014). This alone may discourage them from speaking up in class or seeking out help.

Additionally, first-generation college students often are trying very hard to fit in and not be noticed. They do not want to stand out by asking questions that show that they don't know something (Cox, 2009; Newman, 2000). Their lack of participation in class is frequently to be out of a desire to avoid *taking a risk,* rather than related to how well prepared they are for class (Landers, 2018)

Like the student I described above, whose cell phone died, first-generation college students are often from low income backgrounds. Research shows they are disproportionally so (Lohfink & Paulsen, 2005). They therefore have many additional responsibilities placed on them in comparison to their seemingly STEM savvy classmates. First-generation, low-income college students often work full time while attending college (Engle, 2007; Terenzini, Springer, Yaeger, Pascarella, & Nora, 1995). They may even work at multiple jobs. They may come to school after just having gotten off a full night's shift of work. Sometimes they may be coming to school from a job that is a physically arduous one. While this does not mean it is fine or a good thing if they nod off in your class, it does make for a better explanation than just assuming they were out partying the night before your class.

First-generation, low-income students are more likely to be commuter students and more likely to be living at home with family members (Christensen, 2016; Engle, 2007). While living at home saves money, it frequently comes with the added responsibilities of caring for family members (such as younger siblings or sick, elderly, or disabled family members). These responsibilities often interfere with and impede students' academics.

First-generation, low-income students and their families often live close to the edge financially, in some cases paycheck to paycheck. So that when a crisis happens (e.g., a family member gets sick and cannot work, or a natural disaster hits and harms the family economically), the student and his or her academics are really interfered with (Christensen, 2016).

The various situations and socioeconomic issues that low-income first-generation college students face are not always obvious ones. Often they are hidden, not readily visible and easily overlooked. It is therefore important for you to get to know your students and to ask questions of them. This will enable you to be better informed about them so that when they arrive late, don't turn things in, and so on, you will know whether it is a case of a student slacking off or a situation that is beyond the student's control.

How to handle these situations and how to penalize the student is something you will have to discuss with the faculty and staff at your institution. But knowing the motivation behind students' behaviors is very important. It will affect the messages of encouragement and discouragement that you send to your students. It will also impact your ability to work with students. If you have the actual facts of their situations at hand, you will be able to be a much *more effective* instructor and educator.

Below, I share some specific pointers that will enable you to support, foster, and encourage the success of your first-generation students.

The Strengths of First-Generation Students

Before I discuss how you can encourage first-generation students to adopt some of the behaviors that STEM faculty generally think of as good habits of successful STEM students, I want to clue you in on some of the perhaps not so obvious assets and strengths of first-generation college students.

First-generation college students are extremely hard working, determined, and resilient (Pascarella, Wolniak, Pierson, & Terenzini, 2003; Trevino & DeFreitas, 2014). They typically have had to work extremely hard just to achieve the accomplishment of making it to college. Many of them come from weak or failing high schools, schools that do not have adequate resources (Engle & Tinto, 2008; Terenzini, Springer, Yaeger, Pascarella, & Nora, 1995). Much of their science and math knowledge may be self-taught. Many of them are used to putting in long hours of physical labor at low paying jobs. They likely come from families where hard work is valued and where making ends meet often means taking on another part time job and/or sleeping less.

You, Yes You, Can Help First-Generation Students Be Successful

The resilience, determination, and work ethic of first-generation students is perhaps their secret weapon. But it is up to you to help them tap into it. You can guide them, inspire them, and motivate them.

Encourage Questions

Try to create an atmosphere in your class where all students feel that asking questions is encouraged and welcomed. Avoid saying things like "Is this clear?" or "Is everyone clear?" These types of questions put a lot of pressure on students to avoid answering no. Instead, use phrases like "What about this is not clear?"; "Help me out. How could I make this more clear?"; or "What is the most confusing aspect of this?" Then wait 30 seconds (count silently in your head) until someone can't stand the silence anymore and volunteers a question (Ober & Saltzman, 2017; Tobin, 1987).

Keep in mind that if you ask the class "Is this clear?" and you get no answer from anyone you will know absolutely nothing about whether or not students are confused or what actually is confusing them. Because, even if you get some head nods of yes to your question, you will still not know who doesn't understand or what they do not understand.

Another low risk, nonthreatening way to get students to ask questions is to require all students to anonymously submit questions on index cards (Ober & Saltzman, 2017) or to have students hand in a written description of what their "muddiest point" was for a given class period (Angelo & Cross, 1993).

Encourage Participation, but Low Stakes Participation

Some instructors try to encourage individuals to participate in the answering of instructor posed questions by either calling on students by name or inviting individual students up to the blackboard. I personally avoid asking individuals to come up the blackboard as asking an individual to come up by him or herself puts the student on the spot. I also find that if I ask for volunteers to come up to the blackboard, only very strong or very confident students volunteer.

Neither of these methods encourages the more comprehensive kind of participation that I am looking for. I find that a better way of getting students to participate at the blackboard is to assign problems to small groups of students and then have each *group* get up and put up their solution to a particular problem on the board as a *group*. Another low stakes method of facilitating participation is simply to have students work in groups on problem solving (Newman, 2000). Having students quietly discuss among themselves how to solve a problem or a series of problems allows them to make hypotheses about how to approach the problem, attempt to tackle the problem, make errors, get stuck, and so on, with some guidance from peers, but without having to publicly expose themselves and voice their thoughts to the whole class.

While you have students working in groups, you can walk around the room and listen in on what students are doing and gather some actual facts or evidence. In this way, you can learn a lot about who is prepared, who is trying, who is lost, and so on without making the assumptions you might make just based on how students present themselves when in front of the whole class.

Another way to encourage participation when you are presenting (e.g., when you are up at the blackboard, perhaps working out a problem in front of the whole class) is to ask very simple, (e.g., multiple choice) questions of the class and have students call out the answers. For example, you could ask the class a question of the following type: "Which reagent should I choose here as my attacker, the acid, or the base?" This type of question might seem like a very trivial question. But, it is a *purposefully* trivial question! Some students will know the answer and get it right. However, almost all students will attempt the question even if they don't know the answer, because they know in advance what all the possible answers are. Also, they know that their choices are very limited, so there is a very limited risk involved in guessing.

A multiple choice question with only two choices is a very low risk situation for students because odds are pretty good they will answer the question correct (especially if the question is not a difficult one). So why ask it, then? It is a way to *involve* students and keep them engaged. It is a way to make all students feel they are part of the conversation. This will then help them feel like it is okay to stop you later on and to ask a question as you continue to work through the problem.

I would argue that the converse of asking easy questions, asking vague, open ended questions, such as "What is missing?" or "What is going to happen next?" is often not a good idea. Because such questions are broad, yet have very specific answers, they will not encourage all students to participate. Such questions are more likely to be perceived of by students as challenging, higher risk type situations. Students who want to avoid exposing themselves or embarrassing themselves in front of their peers will avoid attempting these types of questions even if they know the answers (Landers, 2018). These questions may be useful to include at times for educational purposes, but they will not serve the purpose of encouraging broad participation by all students.

Help First-Gen Students Effectively Tap Into Their Work Ethic and Resilience

Often times, I find that first-generation college students will put in tons of effort towards a STEM course, but often not in a particularly strategic or fruitful way. For example, a student might get stuck when trying to solve a particular homework problem and then spend 4 or 5 hours reading the textbook and/or watching online videos trying to learn or understand what he or she needs to know in order to solve the problem. He or she would likely have been better off, from an efficiency standpoint, having spent only a short amount time trying to solve the problem before skipping it and moving on to a different problem, while then perhaps later coming in to office hours or emailing to get help with the problem he or she was stuck on.

I have learned that first-generation college students can be funny that way. They are such hard workers and so stubborn and resilient (Pascarella, et al., 2003; Trevino & DeFreitas, 2014) that they sometimes just keep banging their heads against a brick wall over and over thinking that eventually the wall will come down. And sometimes, after tremendous time and effort, they do make progress through brute force. But often, they waste much valuable time using a brute force approach to studying, rather than by using a more strategic approach. In other chapters of this book, I have discussed how professors can help students to be more strategic (Chapter 5) and how first-generation students themselves (Chapter 9) can learn to be more strategic.

But you have an important role to play in these types of situations as well!

Sometimes it can be difficult to be a graduate student instructor. Students may recognize that you are a student and less experienced than a professor. They may even go so far as to challenge your authority or be disrespectful. But your graduate student status can also be an advantage and asset. You may be close in age to your students. They likely relate better to you than to the course professor because you, like them, are a student (Seymour, Melton, Wiese, & Pedersen-Gallegos, 2005). Students trust that you know how to be a successful STEM student. You are living proof of someone who is similar to them who has successfully made it through an undergraduate STEM major. And you are actually an expert in how to study successfully in the discipline!

When you see a student figuratively banging his or her head against the wall or engaging in behaviors that are likely to harm, rather than help, his or her academic performance, see if the student is open to talking to you about more effective ways to approach his or her studies, whether these be methods you have found to be useful or whether they be methods I describe in Chapter 9.

Encourage Help Seeking

One challenge that first-generation college students face is that they are very reluctant to seek out human help and/or make effective use of campus resources such as advising, tutoring, office hours, and so on (de Cordova & Herzon, 2007; Stephens, Hamedani & Destin, 2014; Nichols & Islas, 2016). First-generation students are accustomed to being self-reliant and dependent only their own family members for survival (Landers, 2018). Seeking help outside the family may go against the social and cultural norms with which they have been raised (Johnson, 2016). Added to this is a desire to remain invisible and not to let on that they are different or in any way in need of help (Landers, 2018). But you as a graduate student, whom they may perceive of as somewhat of a peer, can clue them in to the fact that the best and strongest of students are actually those who *take advantage of resources*, and that if they truly want to fit in and emulate everyone else, they need to start becoming help seekers as well, especially if they want to become successful STEM students.

Note

1. The words university and college are often used interchangeably in the United States. In the United States, colleges are not usually lower ranking institutions than universities, nor are they more likely to offer technical degrees. Colleges are simply institutions that *only* offer bachelor's and master's degrees (e.g., Columbia College), while institutions termed universities must offer at least one doctoral degree (e.g., Harvard University).

References

Angelo, T. A., & Cross, K. P. (1993). *Classroom assessment techniques: A handbook for college teachers* (2nd ed.). San Francisco, CA: Jossey-Bass.

Augenblick, J. G., Myers, J. L., & Anderson, A. B. (1997). Equity and adequacy in school funding. *The Future of Children, 7*(3), 63–78.

Bowen, W. G., Kurzweil, M. A., & Tobin, E. M. (2005). *Equity and excellence in American higher education.* Charlottesville, VA: University of Virginia Press.

Carnevale, A. P., & Strohl, J. (2010). How increasing college access is increasing inequality and what to do about it? In R. D. Kahlenberg (Ed.), *Rewarding strivers: Helping low-income students succeed in college* (pp. 71–183). New York, NY: Century Foundation Press.

Christensen, W. M. (2016, September 9). Advising as activism. *Inside Higher Ed.* Retrieved from https://www.insidehighered.com/advice/2016/09/09/importance-advising-first-generation-students-essay

Cox, R. D. (2009). *The college fear factor: How students and professors misunderstand one another.* Cambridge, MA: Harvard University Press.

Davis, J. (2010). *The first generation student experience.* Sterling, VA: Stylus.

de Cordova, H. G., & Herzon, C. (2007). From diversity to educational equity: A discussion of academic integration and issues facing underprepared

UCSC students. *UC Berkeley: Center for Studies in Higher Education.* Retrieved from https://escholarship.org/uc/item/5sc6j0rx

Engle, J. (2007). Postsecondary access and success for first-generation college students. *American Academic, 3,* 25–48.

Engle, J., & Tinto, V. (2008). *Moving beyond access: College success for low-income, first-generation students.* Washington, DC: The Pell Institute.

Gabriel, K. F. (2018). *Creating the path to success in the classroom: Teaching to close the graduation gap for minority, first-generation, and academically unprepared students.* Sterling, VA: Stylus.

Jackson, M. C., Galvez, G., Landa, I., Buonora, P., & Thoman, D. B. (2016). Science That Matters: The Importance of a Cultural Connection in Underrepresented Students' Science Pursuit. *CBE-Life Sciences Education, 15*(3). doi:10.1187/cbe.16-01-0067

Jehangir, R. R. (2010). *Higher education and first-generation students: Cultivating community, voice, and place for the new majority.* New York, NY: Pellgrave Macmillan.

Johnson, E. (2016, October 30). Micro-barriers loom large for first-generation students. *Chronicle of Higher Education.* Retrieved from https://www.chronicle.com/article/Micro-Barriers-Loom-Large-for/238218

Landers, K. H. (2018). *Postsecondary education for first-generation and low-income students in the Ivy League: Navigating policy and practice.* Cham, Switzerland: Palgrave MacMillan.

Lohfink, M. M., & Paulsen, M. B. (2005). Comparing the determinants of persistence for first-generation and continuing-generation students. *Journal of College Student Development, 46*(4), 409–428.

McGarty, C., Yzerbyt, V. Y., & Spears, R. (2002). Social, cultural and cognitive factors in stereotype formation. In C. McGarty, V. Y. Yzerbyt, & R. Spears (Eds.), *Stereotypes as explanations: The formation of meaningful beliefs about social groups* (pp. 1–15). Cambridge, England: Cambridge University Press.

National Foundation for American Policy. (2017). *NFAP Policy Brief: The importance of international students to American science and engineering.* Retrieved from https://nfap.com/wp-content/uploads/2017/10/The-Importance-of-International-Students.NFAP-Policy-Brief.October-20171.pdf

Newman, R. S. (2000). Social influences on the development of children's adaptive help seeking: The role of parents, teachers, and peers. *Developmental Review, 20,* 350–404.

Nichols, L., & Islas, A. (2016). Pushing and pulling emerging adults through college: College generational status and the influence of parents and others in the first year. *Journal of Adolescent Research, 31*(1), 59–95.

Ober, T., & Saltzman, E. (2017). Achieving total student participation in today's diverse college classes. In R. Obeid, A. T. Schwartz, C. Shane-Simpson, & P. J. Brooks (Eds.), How we teach now: The GSTA guide to student-centered teaching (pp. 107–122). Retrieved from http://teachpsych.org/ebooks/howweteachnow

OECD. (2016). *Education at a glance: OECD indicators.* Paris, France: OECD.

Pascarella, E. T., Wolniak, G. C., Pierson, C. T., & Terenzini, P. (2003). Experiences and outcomes of first-generation students in community colleges. *Journal of College Student Development, 44*(3), 420–429.

Priniski, S. J., Hecht, C. A., & Harackiewicz, J. M. (2018). Making learning personally meaningful: A new framework for relevance research. *Journal of Experimental Education, 86*(1), 1–19.

Rheinschmidt, M. L., & Mendoza-Denton, R. (2014). Social class and academic achievement in college: The interplay of rejection sensitivity and entity beliefs. *Journal of Personality and Social Psychology, 107*(1), 101–121.

Ross, L. (1977). The intuitive psychologist and his shortcomings: Distortions in the attribution process. *Advances in Experimental Social Psychology, 10,* 173–220.

Seymour, E., Melton, G., Wiese, D. J., & Pederson-Gallegos, L. (2005). *Partners in innovation: Teaching assistants in college science courses.* New York, NY: Rowman & Littlefield.

Stephens, N. M., Hamedani, M. G., & Destin, M. (2014). Closing the social-class achievement gap: A difference-education intervention improves first generation students' academic performance and all students' college transition. *Psychological Science, 25*(4), 943–953.

Terenzini, P., Springer, L., Yaeger, P. M., Pascarella, E. T., & Nora, A. (1995, May). *First-generation college students: Characteristics, experiences, and cognitive development.* Paper presented at the Meeting of the Association for Institutional Research, Boston, MA.

Thorne, R. (2013, October 1). 3 ways to help first-gen college students survive in STEM courses [Blog post]. Retrieved from https://sparkaction.org/content/3-ways-first-gen-survive-STEM-college

Tobin, K. G. (1987). The role of wait time in higher cognitive level learning. *Review of Education Research, 57*(1), 69–95.

Trevino, N. N., & DeFreitas, S. C. (2014). The relationship between intrinsic motivation and academic achievement for first generation Latino college students. *Social Psychology of Education, 17,* 293–306.

Trevino, N. N., & DeFreitas, S. C. (2014). The relationship between intrinsic motivation and academic achievement for first generation Latino college students. *Social Psychology of Education, 17,* 293–306.

UNESCO. (n.d.). *World Inequality Database on Education: Upper secondary completion rate.* Retrieved from http://www.education-inequalities.org/indicators/comp_upsec_v2#?sort=mean&dimension=all&group=all&age_group=comp_upsec_2029&countries=all

7

Understanding the Psychological Factors That Foster/Hinder Student Adoption of Self-Regulated Learning Behaviors

The purpose of this chapter is to help you better understand what is going on inside the hearts and minds of your first-generation students so that you can work with them most effectively. This chapter will describe and explore a number of psychological factors that impact first-generation students. Many of these factors impact the feelings and behaviors of students in negative ways and prevent them from adopting or improving their self-regulated learning behaviors. However, the chapter will also demonstrate that not all psychological factors associated with first-generation college students are maladaptive.

The chapter will also describe a variety of psychological interventions that have been conducted with first-generation students and similar populations of students. Such interventions have demonstrated that when students are better able to understand how their identities and backgrounds impact their academic journeys, these attributes can be sources of strength

Teaching STEM to First Generation College Students, pages 67–91
Copyright © 2019 by Information Age Publishing
All rights of reproduction in any form reserved.

and support that lead to academic success. For each intervention described, the chapter will provide pointers for you as to what you can learn, take away, and utilize to help your students.

Imposter Syndrome

Instructors as Imposters

Do you remember what it felt like when you were a new instructor teaching for the first time, either as a new graduate student or new faculty member? Were you nervous the first time you had to get up in front of a classroom full of college age students?

If yes, your feelings were not unusual. Not all instructors get nervous when they teach for the first time, but many do. Many worry that students will quickly ferret out that they are newbies, that students will immediately be able to tell, almost as if it is printed on their faces or clothing, that this is their first time teaching (McManus, 2005). New instructors worry that students will question their competence or even challenge their authority in the classroom.

Most of us plow through our early teaching experiences by "faking it" (McManus, 2005). We pretend to know what we are doing. We act with a confidence we don't feel, until eventually we get acclimated and adjusted, gain skills, experience, and confidence. Basically, we "fake it, until we make it." Eventually, our feelings of being novices or imposters who don't really belong, fade.

First-Generation Students as Imposters

New instructors are not alone in feeling like imposters. First-generation college students often feel and behave like imposters (Davis, 2010, Jehangir, 2010; Peteet, Montgomery, & Weekes, 2015; Stebleton & Soria, 2013).

Imposter Feelings

First-generation college students often feel lost when they enter higher education. They are unfamiliar with the landscape and culture (Davis, 2010). More specifically, they are unfamiliar with the rules and regulations of universities and are unfamiliar with the terminology and jargon in use (Collier & Morgan, 2008). For example, they lack clear understanding of terms like withdrawal, incomplete, corequisite, and so on.

First-generation college students feel anxious about their lack of knowledge and understanding of higher education (Pascarella, Wolniak, Pierson, & Terenzini, 2003). They often question whether or not they belong in college (Rheinschmidt & Mendoza-Denton, 2014), especially when they look around them and observe classmates (e.g., continuing generation students) who seem well-adjusted, academically successful, and socially at ease (Davis, 2010; Gabriel, 2018; Landers, 2018).

When first-generation college students compare themselves to their more successful peers, they often conclude that their disorientation, lack of savvy, struggles, and failures are due to faults and weaknesses within themselves (Cox, 2009). This conclusion leads first-generation students to try to hide their perceived faults and weaknesses from others. In effect, first-generation students are trying to "fake it, til they make it."

Imposter Behaviors

First-generation college students "fake" their way through higher education by trying to maintain as low a profile as possible. They figure if they maintain a low profile, no one will notice them and no one will notice that they don't belong. But more importantly, they figure no one will notice their *shortcomings* and their *mistakes*. By maintaining a low profile, they minimize the risk of being exposed for the failures that they believe themselves to be (Cox, 2009).

Hiding

The types of behaviors that first-generation students engage in in order to maintain a low profile are not surprising. They avoid raising their hands in class to answer questions, lest they be called on. They also avoid asking questions or participating in class discussions (Newman, 2000). While all forms of classroom participation could be opportunities for first-generation students to engage with the material, get to know their peers and instructors, or provide a different perspective on the content, for first-generation students, they are instead instances which risk exposure of their ignorance (Cox, 2009).

Ironically, students who don't participate can come across as unengaged, uninterested, or unprepared. But, as is the case of many first-generation students, they are not at all neutral about their courses (Smith, 2018). They may be very interested in the subject matter, but rather are anxious to remain as invisible as possible (Cox, 2009; Landers, 2018).

One way you can respond to students who are feeling afraid of exposure is to remind them that they are in the same boat as everyone else, that everyone struggles with the material, and that there are particularly difficult topics that trip up most students, even the strongest ones. You can also encourage students to ask questions by having students submit questions in an anonymous way (for example by having them submit an index card at the end of class) where they can write down what their sticking points were or what the murkiest point was for them (Angelo & Cross, 1993). Additionally, responding respectfully to all student questions in class (even if just to let a student know that you will have to get back to him or her after class) sends the message to students that asking questions and looking critically at the material is an important part of learning (Ober & Saltzman, 2017).

Another way to encourage questions from students is to describe for them one or two examples or stories of topics you struggled with when you were in school. Letting students know that you are or were not perfect will help them be more comfortable admitting to themselves and to others that they are fallible as well.

Not Learning From Others

The imposter syndrome experience of first-generation college students creates additional problems. Many first-generation college students are so busy hiding from others (and from themselves) that they can't pay attention to the useful behaviors of the best studiers (best self-regulated learners) around them.

What do I mean by this? Some of my most successful students are those who come to college not knowing how to study effectively, but who are able to learn and improve their study habits by emulating the best studiers (the best self-regulated learners) around them. Some first-generation college students are able to evolve and develop strong self-regulated learning skills by comparing themselves to others and by learning from these comparisons. However, this requires first-generation students to first recognize that their shortcomings should not be sources of shame, but rather are skills that can be developed and improved (Butera & Darnon, 2018).

Avoiding Risk at All Costs

First-generation students don't just manage to avoid noticing and learning from the behaviors of their stronger classmates. They actively try to avoid learning new behaviors. Because they don't want to look stupid in the eyes of others, they are risk (or performance) avoidant (Elliot & Mg-Gregor, 2001). They do not want to take the risk of trying to learn any new behaviors, such as new study behaviors. Learning any new behavior is a situation where the risk of making mistakes can be great. So presenting ideas about how to change study behaviors to first-generation college students can be a challenge as this can be very threatening to them.

Even if what they are doing study wise is not working, first-generation students may continue to do it over and over and over, rather than taking a risk to try something new (Caselman, Self, & Self, 2006). For example, they may repeatedly sign up for and withdraw from or fail a course that they are unprepared for (to the detriment of their college transcripts) or repeat the same improper, inferior study strategies that have not worked for them over and over again in a manner sometimes referred to as functional fixedness (Duncker, 1945).

Fixed and Growth Mindsets

The STEM Brain?

When I meet strangers in social situations, I dread their inevitable question asking what I do for a living. I dread the question, because my answer, "I teach chemistry," is a certain conversation stopper. The most common responses I get from folks when I tell them that I teach chemistry are, "I failed chemistry in high school"; "Chemistry was my worst subject"; or "Wow, you must be *so* smart." Immediately these strangers, who are often bright, accomplished professionals in their own fields, set themselves apart from me as if I am some unusual, almost abnormal specimen, leaving me feeling awkward and unsure as to how to respond to their comments.

But it is clear from the comments that I get and from those that other STEM faculty report receiving (McGrew, 1993) that there is a prevalent attitude in our society that only some, "special" folks can understand and do science and mathematics, that only some folks are privileged to have the "right" type of brains, science brains, or mathematics brains.

Fixed and Growth Mindsets

STEM faculty hear this attitude or mindset from students as well. Students will say things like "I can't do math"; I'm no good at biology"; or "I'm just not a science person." The idea that one's ability in STEM (or in other academic and even nonacademic areas) is fixed and not changeable, for example, through effort and hard work, has been termed a *fixed mindset* by well-known social psychologist Carol Dweck (2006). In contrast, if one believes that one's ability (e.g., academic ability) can be improved through effort and practice, Dweck terms this mindset a *growth mindset.*

Academic Consequences of Fixed and Growth Mindsets

It might not surprise you that students' beliefs regarding how malleable, changeable, or controllable their academic abilities are impact their academic performance. Research demonstrates that students with growth mindsets earn higher grades than students with fixed mindsets (Dweck & Molden, 2017). Students with growth mindsets also respond differently to failure. They are more likely to plan to work harder and/or to put in more time, while those with fixed mindsets are more likely to plan to devote less time to studying or to give up altogether. Students with growth mindsets report approaching academic challenges and difficulties by trying harder, while fixed mindset students report feeling discouragement and contemplating giving up (Dweck & Molden, 2017).

First-Generation Students: Self-Doubt and Self-Blame

The consequences of having a fixed mindset are particularly problematic for first-generation college students who can arrive at college feeling like imposters and who may arrive with poorer math and science backgrounds and with poorer study habits. But rather than realizing that their STEM backgrounds and study habits are due to environmental factors, first-generation students immediately start to doubt themselves and often rather quickly come to the conclusion that their poor performance is due to their own inferiority. To put it another way, first-generation college students often come to college thinking that they are smart and good at STEM, but they then quickly conclude that they are wrong when they begin to compare themselves to their better prepared continuing-generation peers (Davis, 2010; Landers, 2018).

Lack of Self-Regulation

Soon the self-doubts of first-generation students become feelings of self-blame. And when first-generation students start to blame themselves for their failures, the opportunity for them to adjust and improve their academic self-regulation gets lost. Students who believe their brains are "just not good enough," start to give up. Instead of thinking about strategy usage, these students starting actively avoiding experiences that remind them of what failures they are (Dweck & Master, 2008; Greene, Costa, Robertson,

Pan, & Deekens, 2010; Stump, Husman, & Corby, 2014). They stop coming to class. They avoid seeking out their professors and they don't bother to avail themselves of campus resources or sources of help like on campus tutors. Ironically, such students often come across as unengaged and unmotivated, when in fact these students care very much about their performance and may be feeling very upset and anxious about it (Cox, 2009).

Parents of First-Generation Students

The mindsets of parents also play a role in the academic performance of first-generation STEM students. Often when first-generation STEM students struggle in their STEM courses, they receive messages and direct instructions from their parents as to how to modify their academic behaviors and study habits. Their parents, however, are typically not well versed in the world of higher education, and are not able to differentiate between different types of STEM courses. They therefore cannot guide their children as to how one might study differently in one STEM course versus another (Nichols & Islas, 2015).

Many parents of first-generation students who see their children struggle in these situations simply tell their children what they know from their own life experiences—which is to work harder (Nichols & Islas, 2015). These parents are often used to struggling financially and are familiar with having to add more work hours, take on more jobs, or go without sleep in order to survive. So this is what they tell their children to do. But this advice is typically not useful because usually their children need to study smarter, not harder. The children need to be more savvy and strategic, and need to learn to use their time more effectively (Davis, 2010).

Furthermore, when parents who advise their children to work harder observe that their children have increased their efforts and amounts of time and are still not succeeding, these parents then conclude, in a fixed mindset way of thinking, that their children are just not cut out for science, that they are just not good enough, or not smart enough. So when these parents see that the extra effort is not paying off for their children, the parents tell their children, well I guess you just are not cut out for this. You really should consider doing something else (Tough, 2014).

The fixed mindset about STEM performance that parents have can be very damaging to first-generation students (Pomeranz & Grolnick, 2017). The students feel like failures. They feel like they have failed their parents and like their parents have given up on them and stopped believing in them. But the parents also reinforce for them their own individual beliefs that they are not good enough and that *their efforts do not matter.* This is very damaging in terms of trying to get students to improve their study habits because it requires students to overcome the idea that their own parents do not believe they are capable of achieving in STEM, before they can begin to try to improve or change their behaviors. Students have learned from birth to trust their parents, so despite the fact that higher education is not their parents' area of expertise, it is a big blow to students when their parents do not believe they are capable.

Further Consequences of Having a Fixed Mindset

When students believe that they are the ones at fault, that they are the ones to blame for their own performance, and that there is nothing they can do about their performance, they adopt a number of strategies to avoid the negative feelings that come along with feeling like failures. These strategies are often detrimental to students, particularly in terms of their being able to improve their self-regulated learning abilities.

Avoidance Behaviors

Sometimes when students feel awful about how they are doing in a course (or in a number of courses), they will adopt behaviors that enable them to avoid experiencing their feelings. Particularly, they will avoid situations where they *experience failure* itself and they will avoid situations where they *experience feedback* indicating that they are failing (Cox, 2009).

Avoidance of Evidence

Ever notice that poorly performing students are the ones who miss the most classes or who more frequently come late to class? While it is true that some students might simply perform poorly because they have missed significant amounts of important class material, other students are deliberately (although not consciously) not coming to class or deliberately not prioritizing coming to class. Why? When they come to class, they feel lost and confused. And then, they also observe other students around them who seem to get it, who are on top of the material. They observe students who seem to absorb the material quickly and easily. Coming to class provides them with evidence and reminds them of their own gut sense that they are not cut out for this. So they make excuses (to you and to themselves) for why they could not make it to class or for why they were late (Nurmi, Aunola, Salmela-Aro, & Lindroos, 2003).

Self-Handicapping

Similarly, students who want to avoid evidence that they are in trouble, avoid taking quizzes and exams. They miss them altogether, or arrive late, thereby sabotaging their chances of performing decently (Gabriel, 2018; Grant & Dweck, 2003; Zimmerman & Schunk, 2008). Once they have sabotaged their chances, they have a good rationale that they can offer to themselves as to why they didn't do as well as they could have. They comfort themselves by telling themselves that they really did know the material, but unfortunately because the bus was running late or they had trouble finding parking, they did not get to demonstrate what they actually knew (Cox, 2009).

Procrastination

You might think that procrastination is simply a laziness, passiveness, or lack of behavior on the part of students. But, I view procrastination as a deliberate (although not conscious) action or symptom. Procrastination is another strategy that students use to sabotage themselves (Dweck & Molden, 2017), by delaying and putting off school work and not studying until the last minute. When procrastinators do poorly, they can rationalize to themselves that they really did know the material and that they will of course do better the next time when they have more time to study, that this time it was just an unfortunate accident where they simply couldn't do their best because they didn't have enough time.

Procrastinating and self-handicapping students are doing everything they can to avoid facing the truth. Consequently, they adopt habits and behaviors that are exactly the opposite of what STEM faculty would consider ideal habits for success, such as coming to exams early and making sure to get a good seat, studying a little bit everyday, and so on.

Avoidance of Feedback

You may have noticed that the same students who have poor attendance habits (and poor performance) also tend to not pick up their graded quizzes and exams. Is this just sloppy or lax behavior on their part? Maybe in some instances. But, often it is an *active* avoidance of feedback. Not picking up one's graded work (or picking it up, but immediately stashing it somewhere out of sight, never to be found again) is often a deliberate attempt on the part of a poorly performing student to remain in denial, to avoid confronting evidence of failure (Cox, 2009).

This avoidance behavior is unfortunate. Students with avoidance behaviors have no way of learning from their mistakes. They do not review their errors. They do not review the correct answers. They do not learn which parts of the material are less or more important. They do not figure out what their strengths and weaknesses are. They do not learn how to better tackle exams or how to gauge how to study more effectively for the future. Much self-regulation comes from assessing one's performance (based on feedback) and then modifying one's behavior accordingly (Zimmerman, 1990). This self-assessment is completely missed out on by students who cannot confront evidence of their perceived shortcomings.

You probably would not be surprised that students who want to avoid written feedback also avoid your office hours like the plague. Talking to you (inside or outside of class) is fraught with danger for them. They are not uncaring or unengaged. However, they are terrified that you will see right through them and see them for the true failures that they believe themselves to be (Cox, 2009). They, of course, lose out on your valuable experience by avoiding you. You very likely have some very good ideas of what they can do (think: growth mindset) to improve. You will also likely be able to very quickly, by looking at their exams, diagnose their areas of deficiency and guide them as to where to place their focus and emphasis when trying to remediate themselves and/or get caught up.

Ironically, if avoider type students understood that their failings were indeed malleable and were often simply due to less developed study skills, they could potentially improve. They would then also feel much less self-blame and shame regarding their performance. Unfortunately however, the shame/blame and fixed mindset situation can be kind of like a vicious cycle that prevents change and improvement in academic self-regulation, especially on the part of first-generation students.

Quitting

A more drastic step that students may take to avoid negative feedback about their abilities is to give up or quit altogether. This may take the form of dropping a course, switching out of a STEM major, or even dropping out of college altogether (Cox, 2009). While switching out of something in which one is unsuccessful might seem like a strategic and adaptive strategy, if one simply switches to avoid the potential experience of failure (which is sometimes a necessary component of learning), many students who simply lack the proper study skills may end up giving up on STEM and/or on higher education for rather faulty reasons.

When It's Not the Student's Fault

Students are pretty good (even if it is not conscious on their part) at avoiding feelings of failure, because these feelings are very powerful and come along with similarly powerful feelings of shame. So they avoid, deny, sidestep, and procrastinate, to avoid confronting evidence that *they* are at fault.

And sometimes, they try another tactic: They find ways to blame others.

Avoidance of Responsibility/Blaming of Others

One way of not feeling like a failure is to point the finger at other folks (Weiner, 1985). Students who are in serious trouble may try to blame you. They may blame you for doing a poor job in your teaching, for writing a lousy or unfair exam, for grading unfairly, or even, for discriminating against them. Underneath their bluster and anger (particularly if they are a first-generation student) is usually a very upset, scared, and struggling student who is looking for someone to lash out at.

The deflecting of responsibility onto you may seem to students like a very effective strategy. You may get defensive. Even if you don't, their combative attitude may distract you from examining or talking to them about their performance. Maybe you will even find a small error in the grading and you will return a couple of points to the student so that momentarily the student will feel like he or she is doing better. But attacking you, will likely just distract both you and the student from the true problem of the struggling and lost feelings that the student is experiencing.

Responding to the student's challenges and attacks in a manner that legitimizes them is problematic because it supports the idea (to the student) that he or she is not responsible for his or her grades, rather that you are. Getting the student to understand that he or she is in control of his or her performance is crucial, if the student is going to buy-in that modifications to study habits are needed and necessary.

Learned Helplessness

Sometimes when students have experienced repeated failures or repeated feelings of hopelessness (that nothing they can do will enable them to be successful), they may look to blame you or others in their lives. But in some cases, they may develop a psychological condition known as learned helplessness (Abramson, Seligman, & Teasdale, 1978) in which they feel helpless and hopeless, have given up, and are no longer trying. Because they feel that nothing that they do matters or makes a difference, they stop trying and stop putting in effort and of course stop thinking about what strategies are working, not working, and/or should be tweaked.

Learned helplessness can be unlearned when students are presented with evidence contrary to their beliefs, namely that with proper effort, they can be successful. Evidence that their performance is malleable and may be in fact due to their socioeconomic backgrounds can be useful in helping convince students as to why they are in the difficult academic situations they find themselves in (Stephens, Hamedani, & Destin, 2014).

Academic Help Seeking

Have you ever noticed that some of your strongest students are the most proactive ones? They are the ones who seek you out with questions during your office hours and/or who come up to you before and after class with questions. They may even seek you out in your office outside of your office hours. They very likely also email you with questions. In the language of psychology and self-regulated learning, these students are *help-seekers*.

Help-seeking is a very important part of learning, an important part of self-regulated learning. Recognizing that one needs help, deciding to seek out help, and choosing the appropriate sources of help are all parts of the help seeking process that students must learn as they progress through school (Karabenick & Berger, 2013).

Even more so in STEM. Seeking help when one is stuck, lost, or confused is a key skill that STEM students need to master in order to successfully navigate through complex course material. Asking questions, learning from mistakes, finding sources of information, and utilizing resources, are all crucial skills that STEM students need to order to survive and be successful (Karabenick, 2003; Nandagopal & Ericsson, 2012).

Obstacles to Help Seeking

But seeking help is not always a simple process for students. Some students feel confident and comfortable asking for help and admitting when they don't know things, but for many students asking for help can feel risky and threatening, and exposing and embarrassing (Karabenick & Berger, 2013). Students may avoid asking for help because they don't want to look stupid in front of their peers and/or in front of faculty (Collins & Sims, 2006; Newman, 2000; Newman, 2008).

First-Generation Students and Help Seeking

Research shows that first-generation college students are less likely to seek academic help as compared to continuing generation students (de Cordova & Herzon, 2007; Schwartz, Kanchewa, Rhodes, Culter, & Cunningham, 2016; Stephens et al., 2014) and that they are less likely to seek out office hours or form relationships with professors (Cox, 2009; Landers, 2018; Nichols & Islas, 2016).

Imposter Syndrome

For first-generation college students, the barriers to seeking help are great. As was discussed above, when first-generation students feel like imposters, the last thing they want to do is to reveal to themselves and to others (peers and professors) their deep, dark secret that they really don't belong in college (Landers, 2018). Asking questions and seeking help are risky behaviors for first-generation students because these behaviors publicly demonstrate to others that they are lost or confused and that they are not academically strong enough and therefore don't belong (de Cordova & Hervon, 2007; Landers, 2018; Winograd & Rust, 2014).

Lack of Self-Confidence About Higher Education

Additionally, first-generation college students often don't feel the same self-confidence or sense of entitlement within the world of high education that continuing generation students feel (Nichols & Islas, 2016). They often don't feel that they have permission to ask for what they need, to pose questions, or to obtain the help they need (Christensen, 2016; de Cordova & Herzon, 2007). First-generation students often have no one in their family to ask questions of, or no one who can answer the questions they have about the world of higher education (Capo Crucet, 2015; Davis, 2010). First-generation students learn through trial and error that they are going to have to figure things out all on their own if they are going to successfully navigate their academic journeys (Gardner & Holley 2011). The confidence that first-generation students lack prevents them from engaging in academic help seeking.

Lack of Awareness About Higher Education

Ironically, their own lack of awareness about higher education gets in their way and prevents them from obtaining help when they need it. While continuing-generation students are more likely to know about programs and resources available on campus, such as tutoring, academic advisement, and personal counseling (Nichols & Islas, 2016), first-generation college students often don't know that these resources even exist (Winograd & Rust, 2014) or that they are potentially fruitful sources of knowledgeable and useful information (Torres, Reiser, LePeau, Davis, & Ruder, 2006). It is not uncommon for first-generation students to come to college not know-ing what office hours or course syllabi are (Collier & Morgan, 2008). The lack of knowledge that first-generation students have about college and the typical resources one finds at a college makes these students less likely to utilize college resources and less likely to get help when they are in serious need of it (Nichols & Islas, 2016).

Culture of Self-Reliance

Family culture, ethnicity, and socioeconomics also play a role in first-generation students' lack of help seeking. Students who come from im-migrant families and/or working class families often grow up with a very strong ethic of self-reliance, where they learn to rely only on themselves or their immediate family (Davis, 2010; Landers, 2018). Self-reliance may be a matter of pride for a particular first-generation student and his family (Christensen, 2016; Landers, 2018) and asking for help outside the family may be perceived of as an act of serious disloyalty to the family (Johnson, 2016). Additionally, working class, first-generation students may grow up learning that outsiders should not be trusted and that building trust with an outsider, like a college professor or other higher education official, is something that must develop slowly over time (Johnson, 2016; Torres et al., 2006). A distrust of outsiders and culture of self-reliance may discourage or prevent first-generation college students from obtaining academic help in university settings.

Workplace Culture of Compliance/Submissiveness

First-generation college students who grew up in working class house-holds may have learned the hard way that it was normally not a good idea to ask questions of persons in authority. They may have observed that when family members, co-workers, or they themselves asked questions of bosses in the workplace that often questions were not met with acceptance and that the asking of questions in a blue collar environment could risk one's job or antagonize one's boss. First-generation students may have learned that if one did not know what to do on the job that it was sometimes better to keep this information to oneself. First-generation students therefore may have learned that asking questions was not the way to go. However, first-generation students in college don't know the rules of the academic world where asking questions is a valued behavior and a fruitful way of learning (Oldfield, 2007). You need to clue them in about this! Their workplace experiences instead may lead them to exhibit behaviors not conductive to success in college (Stephens, Markus, Fryberg, Johnson, & Covarrubias, 2012).

Strengths of First-Generation Students

Determination

One attribute that I have personally witnessed over and over again on the part of first-generation college students is their incredibly high degree of determination and persistence. Because many first-generation college students come from low-income families (Engle & Tinto, 2008), first-generation students are often accustomed to having had to work for pay from a young age. Often their work is physically grueling (e.g., demolition, line cook), pays poorly, and requires long hours on one's feet. First-generation students who work at low-wage, labor intensive types of jobs think nothing of coming to school after working a full day's shift, or they sometimes just head home for a quick hour nap and shower before starting their school day.

Students who come from families without college educated parents are also accustomed to having parents who work two or three jobs and to parents who have to go without sleep to make ends meet. These students have seen their parents give up food and sleep when necessary. Thus, when a family crisis arises or when it is crunch time (e.g., finals), these students think nothing of sacrificing the basics (such as food and sleep) to power through their studying and coursework.

The willpower, determination, and degree of effort that first-generation students will devote is something that cannot often be matched by continuing generation students. However, such students need to be taught to approach their studies more strategically and less by brute force.

Resilience

Another attribute of first-generation students is their resilience (O'Neal et al., 2016; Pascarella et al., 2003; Trevino & DeFreitas, 2014). First-generation students do not give up easily. They are used to challenges and used to defeats. They may come from very difficult and challenging family situations, yet they manage to overcome these situations and to show up to your classrooms. They refuse to accept defeat. This never give-up attitude means that first-generation students do not give up on their dreams where "weaker" willed folks might move on. This enables some first-generation students to achieve goals and dreams that some may tell them they cannot achieve. However, again, sometimes their dogged determination needs to be channeled towards the most fruitful, strategic study strategies (Smallets, Townsend, & Stephens, 2018).

Research and Interventions Involving First-Generation Students

Where to Go From Here?

Having read all about the psychological barriers that first-generation college students face, you may be feeling somewhat discouraged about how you can possibly help them succeed or overcome these barriers.

Don't be discouraged. Your awareness of the barriers they face is very important and can make a big difference.

For example, knowing that many first-generation students are reluctant to speak up in class or to show up to office hours may stimulate you to change how you do things in your classroom, so that you can provide an atmosphere that is more encouraging and supportive of first-generation students. Another action you can take is to push harder to make sure all students in your classes are well informed about the resources that are available which are associated with your courses. Hopefully, the discussion above will stimulate your thinking and you will come up with your own ideas about how to work more effectively given the psychological needs of first-generation students.

However, the upcoming section of this chapter aims to do more than just give you a pep talk or a chance to reflect and come up with ideas. Below, I describe and review interventional research studies that have been conducted with first-generation college students that have been particularly effective. I also describe why these studies have been effective.

Much can be learned from the rationales and study designs behind these studies. Not that I expect you to go conduct this type of research yourself, but rather I hope that what wisdom has been obtained from these studies will help you when you interact with first-generation college students. In particular, I hope that it will help you frame your conversations with first-generation college students.

Interventions

Social-Belonging Interventions

Research demonstrates that when students feel that they don't belong in a particular social setting, this awareness and feeling is demotivating to them (Walton & Brady, 2017). Yet when students feel a sense of belonging, they are more likely to persist at academic tasks. Even more so, when students perceive that the struggles they are experiencing are normal, everyday struggles that all students go through, they are more likely to feel that they belong in college and to be motivated to persist (Walton & Brady, 2017).

Walton & Cohen (2007, 2011) utilized this theoretical idea to design several *social-belonging* interventions that targeted first-year college students from racial minority groups. Similarly, Yeager et al. (2016) conducted social-belonging interventions with racial minority college students and first-generation college students. In each of these interventions, students were taught that challenges they might experience during their first year of college, such as feeling like they didn't fit in socially, were typical and normal experiences that all college freshman students go through. Students who participated in these interventions were found to be more likely to persist in college; they also earned higher GPAs than nonintervention students (Walton & Cohen, 2011; Yeager et al., 2016). Additionally, participants were found to be more likely to email professors and attend office hours (Walton & Cohen, 2007) and reported feeling a greater sense of belonging (Walton & Cohen, 2011; Yeager et al., 2016).

What can you learn from this? When you talk to students who are struggling, academically or otherwise, it can be very helpful if you let them know that what they are going through is normal, commonplace, and not unique to them. Let them know that adjusting to college level STEM coursework is challenging for all students and that adjusting from high school (where things may have been easier) to college can be a difficult transition for many students. Assure them that even though some students around them look like they are having an easy adjustment, you as the instructor, know that in reality lots of students are struggling.

Sharing this perspective with your first-generation students may help them to feel less shame about their struggles. It may encourage them to open up (with you or elsewhere) about the struggles they are having. It may encourage them to seek help (e.g., tutoring) if this is what they need.

Values Affirmation Intervention

Another approach towards helping first-generation college students, a *values affirmation* intervention, was conducted by Harackiewicz and colleagues (2014). In this method, first-generation college students taking general biology had the opportunity to write about and affirm the core values most important to them. The intervention improved the course performance of the first-generation students significantly. It increased the frequency of their enrollment in subsequent biology courses (Harackiewitz et al., 2014) and enhanced their grades in subsequent courses (Tibbetts, Harackiewicz, Priniski, & Canning, 2016).

What can you learn from this? The *values affirmation* intervention was similar to the *social-belonging* interventions in that it attempted to make students feel that they belonged. However it did so by reminding them of who their core selves really were. When you talk to students, you can ask them questions about themselves, such as, why they came to college, and why they chose their particular major or area of study. Often, first-generation students are particularly interested in STEM or health care for altruistic reasons (Jackson, Galvez, Landa, Buonara, & Thoman, 2016; Priniski, Hecht, & Harackiewicz, 2018). They seek professions in which they feel they can give back directly to their communities or in some way improve the life experiences of folks in their communities. Conversations with students about where they have come from and how their academic and career goals connect back to where they came from can help them feel that there is an academic place, or home for them, at your institution, where their values will be respected and their goals will be recognized.

Education-Difference Intervention

In an *education-difference* intervention conducted by Stephens et al. (2014), first-generation college students were exposed to the idea that their experiences in college might be different (both in negative and positive ways) due to their parents' educational backgrounds and that these backgrounds could be both sources of weaknesses and strengths for them during their college journeys. As a result of this intervention, the first-generation students were more proactive in their use of campus resources and improved significantly in their overall GPAs (Stephens et al., 2014).

What can you learn from this? If your students have overcome personal hardships, whether these be physical, financial, medical, and so on, remind them of what they have survived and overcome. If it is relevant, remind them of the journey they had to go through to get from their country of origin to the United States or the journey they had to undergo to get themselves to college. If they work long hours or work at a job that is physically demanding, remind them of the energy, drive, and physical stamina that they possess.

Reminding students of the strengths that they have, the resilience that they have, and their ability to overcome obstacles, can fortify them to face future obstacles. Letting them know that their traits of a high degree of resilience and endurance may not be common among more economically privileged students can help fortify them when they feel out of place or wonder how they can possibly compete with more economically advantaged peers.

You can also make them aware of the fact that their work ethic and physical stamina can give them an advantage over their more privileged peers, both because they will have greater endurance, but also because potential graduate schools and employers will desire these traits (Horowitz, 2017).

Growth Mindset Interventions

Dweck and others have conducted many studies (see reviews in Burnette, O'Boyle, VanEpps, Pollack, & Finkel, 2013 and Yeager & Dweck, 2012) demonstrating that training students of all educational levels that the "brain is a muscle" that can be developed, enhanced and improved through effort (in other words training students to adopt a growth mindset) will result in significant improvements in their persistence, retention and achievement. In a recent, large scale *growth mindset intervention* conducted with economically disadvantaged college students, Yeager and colleagues (2016) demonstrated that students who received this training were significantly more likely to persist in college and to earn higher GPAs than students who did not receive the training.

What can you learn from this? It can be very challenging to deal with students who truly believe they are "no good at math" or that they "just can't do chemistry." These types of beliefs get in the way of having students hear your messages about what are the most effective ways to study for your course. Sharing a bit with students about what research says and shows, namely that there is not just one type of science or math brain, but rather that it takes a good deal of effort and life experience to become proficient and skilled in a STEM subject, can be a good conversation starter.

But what works best is concrete evidence from you, rather than examples from the research literature. Citing examples from your own experiences, of challenges and struggles that you faced as a student, can be very helpful. Alternatively, you can give examples based on the experiences of students you are familiar with whom you have had recently.

Construal Interventions

The interventions described above were all quite short, yet very effective, interventions. Some were found to have long lasting effects on the student participants, for example, effects on retention, GPAs, feelings of belonging, that were measurable years later.

In an article reviewing different interventions that targeted the achievement gap between lower and higher socioeconomic status students, social psychologists Dittmann and Stephens (2017) grouped together the various interventions I have described (along with a few others) under one umbrella, referring to them as *construal* interventions. In using the term construal interventions, Dittmann and Stephens refer to the fact that these interventions all have in common one aspect, that they target students' construals of themselves, specifically students' understandings or perceptions of themselves within the academic and social environments in which they find themselves. Dittman and Stephens (2017) characterize students' construals as the "processes of meaning making" that students go through in trying to understand and "make sense of" their experiences.

Dittmann and Stephen argue that what makes construal interventions, in particular, so effective, for example, with low socioeconomic status and first-generation college students, is that when students can frame or reframe their experiences as normal, routine and/or sources of strength, rather than as innate flaws (e.g., in their intelligence, in the ability of their racial or ethnic group), they can overcome challenges and obstacles.

What can you learn from this? How can you help students reconstrue their worldviews in ways that can help them be more adaptive and academically successful? You can help students understand some of the rockiness of the journeys that they are on, that adjusting to the college environment does take fine-tuning, that bumps in the road are normal, and that attending college is a transition for students who do not have parents who have gone to or graduated from college. Let your students know that there are many paths into and through college, that it is not true that those sitting around them know all the secrets (they are likely just better at hiding their anxieties), and that fumbling a bit in order to find your way is okay, normal, and maybe sometimes even a good thing.

References

Abramson, L. Y., Seligman, M. E., & Teasdale, J. D. (1978). Learned helplessness in humans: Critique and reformulation. *Journal of Abnormal Psychology, 87*(1), 49–74.

Angelo, T. A., & Cross, K. P. (1993). *Classroom assessment techniques: A handbook for college teachers* (2nd ed.). San Francisco, CA: Jossey-Bass.

Burnette, J. L., O'Boyle, E. H., VanEpps, E. M., Pollack, J. M., & Finkel, E. J. (2013). Mind-sets matter: A meta-analytic review of implicit theories and self-regulation. *Psychological Bulletin, 139*(3), 655–701.

Butera, F., & Darnon, C. (2017). Competence assessment, social comparison, and conflict regulation. In A. Elliot, C. Dweck, & D. S. Yeager (Eds.), *Handbook of competence and motivation: Theory and application* (pp. 192–213). New York, NY: Guilford Press.

Capo Crucet, J. (2015, August 22). Taking my parents to college. *New York Times,* p. SR6.

Caselman, T. D., Self, P. A., & Self, A. L. (2006). Adolescent attributes contributing to the imposter phenomenon. *Journal of Adolescence, 29*, 395–405.

Christensen, W. M. (2016, September 9). Advising as activism. *Inside Higher Ed.* Retrieved from https://www.insidehighered.com/advice/2016/09/09/importance-advising-first-generation-students-essay

Collier, P. J., & Morgan, D. L. (2008). "Is that paper really due today?": Differences in first-generation and traditional college students' understandings of faculty expectations. *Higher Education, 55*, 425–446.

Collins, W., & Sims, B. C. (2006). Help seeking in higher education academic support services. In S. A. Karabenick & R. S. Newman (Eds.), *Help seeking in academic settings: Goals, groups, and contexts* (pp. 203–223). Mahwah, NJ: Erlbaum.

Cox, R. D. (2009). *The college fear factor: How students and professors misunderstand one another.* Cambridge, MA: Harvard University Press.

Davis, J. (2010). *The first generation student experience.* Sterling, VA: Stylus.

de Cordova, H. G., & Herzon, C. (2007). From diversity to educational equity: A discussion of academic integration and issues facing underprepared UCSC students. *UC Berkeley: Center for Studies in Higher Education.* Retrieved from https://escholarship.org/uc/item/5sc6j0rx

Dittmann, A., & Stephens, N. M. (2017). Interventions aimed at closing the social class achievement gap: Changing individuals, structures, and construals. *Current opinion in psychology, 18*, 111–116.

Dweck, C. (2006). *Mindset.* New York, NY: Random House.

Dweck, C., & Master, A. (2008). Self-theories motivate self-regulated learning. In D. H. Schunk & B. Zimmerman (Eds.), *Motivation and self-regulated learning: Theory, research, and applications* (pp. 31–51). Mahwah, NJ: Erlbaum.

Dweck, C., & Molden, D. C. (2017). Mindsets: Their impact on competence motivation and acquistion. In A. Elliot, C. Dweck, & D. S. Yeager (Eds.), *Handbook of competence and motivation: Theory and application* (pp. 135–154). New York, NY: Guiliford Press.

Duncker, K. (1945). On problem-solving (L. S. Lees, Trans.). *Psychological Monographs, 58*(5), i–113.

Elliot, A., & McGregor, H. A. (2001). A 2 x 2 achievement goal framework. *Journal of Personality and Social Psychology, 80*(3), 501–519.

Engle, J., & Tinto, V. (2008). *Moving beyond access: College success for low-income, first-generation students.* Washington, DC: The Pell Institute.

Gabriel, K. F. (2018). *Creating the path to success in the classroom: Teaching to close the graduation gap for minority, first-generation, and academically unprepared students.* Sterling, VA: Stylus.

Gardner, S. K., & Holley, K. A. (2011). "Those invisible barriers are real": The progression of first-generation students through doctoral education. *Equity and Excellence in Education, 44*(1), 77–92.

Grant, H., & Dweck, C. (2003). Clarifying achievement goals and their impact. *Journal of Personality and Social Psychology, 85*(3), 541–553.

Greene, J. A., Costa, L.-J., Robertson, J., Pan, V., & Deekens, V. M. (2010). Exploring relations among college students' prior knowledge, implicit theories of intelligence, and self-regulated learning in a hypermedia environment. *Computers and Education, 55,* 1027–1043.

Harackiewicz, J. M., Canning, E. A., Tibbetts, Y., Giffens, C. J., Blair, S. S., Rouse, D. I., & Hyde, J. S. (2014). Closing the social class achievement gap for first-generation students in undergraduate biology. *Journal of Educational Psychology, 100*(2), 375–389.

Horowitz, G. (2017). First generation college students: How to recognize them and be their ally and advocate. *Journal of College Science Teaching. 46(6),* 6–7.

Jackson, M. C., Galvez, G., Landa, I., Buonora, P., & Thoman, D. B. (2016). Science that matters: The importance of a cultural connection in underrepresented students' science pursuit. *CBE-Life Sciences Education, 15*(3). doi:10.1187/cbe.16-01-0067

Jehangir, R. R. (2010). *Higher education and first-generation students: Cultivating community, voice, and place for the new majority.* New York, NY: Pellgrave Macmillan.

Johnson, E. (2016, October 30). Micro-barriers loom large for first-generation students. *Chronicle of Higher Education.* Retrieved from https://www.chronicle.com/article/Micro-Barriers-Loom-Large-for/238218

Karabenick, S. A. (2003). Seeking help in large college classes: A person-centered approach. *Contemporary Educational Psychology, 28*(1), 37–58.

Karabenick, S. A., & Berger, J.-L. (2013). Help seeking as a self-regulated learning strategy. In H. Bembenutty, T. J. Cleary, & A. Kitsantas (Eds.), *Applications of self-regulated learning across diverse disciplines* (pp. 237–262). Charlotte, NC: Information Age.

Landers, K. H. (2018). *Postsecondary education for first-generation and low-income students in the ivy league: Navigating policy and practice.* Cham, Switzerland: Palgrave MacMillan.

McGrew, L. A. (1993). A 60-second course in organic chemistry. *Journal of Chemical Education, 70*(7), 543–544.

McManus, D. A. (2005). *Leaving the lectern: Cooperative learning and the critical first days of students working in groups.* Bolton, MA: Anker.

Nandagopal, K., & Ericsson, K. A. (2012). An expert performance approach to the study of individual differences in self-regulated learning activities in upper-level college students. *Learning and Individual Differences, 22,* 597–609.

Newman, R. S. (2000). Social influences on the development of children's adaptive help seeking: The role of parents, teachers, and peers. *Developmental Review, 20,* 350–404.

Newman, R. S. (2008). The motivational role of adaptive help seeking in self-regulated learning. In D. H. Schunk & B. Zimmerman (Eds.), *Motivation and self-regulated learning: Theory, research, and applications* (pp. 315–337). Mahwah, NJ: Erlbaum.

Nichols, L., & Islas, A. (2016). Pushing and pulling emerging adults through college: College generational status and the influence of parents and others in the first year. *Journal of Adolescent Research, 31*(1), 59–95.

Nurmi, J., Aunola, K., Salmela-Aro, K., & Lindroos, M. (2003). The role of success expectation and task-avoidance in academic performance and satisfaction: Three studies on antecedents, consequences and correlates. *Contemporary Educational Psychology. 28*, 59–90.

Ober, T., & Saltzman, E. (2017). Achieving total student participation in today's diverse college classes. In R. Obeid, A. T. Schwartz, C. Shane-Simpson, & P. J. Brooks (Eds.), How we teach now: The GSTA guide to student-centered teaching (pp. 107–122). Retrieved from http://teachpsych.org/ebooks/howweteachnow

Oldfield, K. (2007). Humble and hopeful: Welcoming first-generation poor and working-class students to college. *About Campus,* January-February, 2–12.

O'Neal, C. R., Espino, M. M., Goldthrite, A., Morin, M. F., Weston, L., Hernandez, P., & Fuhrmann, A. (2016). Grit under duress: Stress, strengths, and academic success among non-citizen and citizen Latina/o first-generation college students. *Hispanic Journal of Behavioral Sciences, 38*(4), 446–466.

Pascarella, E. T., Wolniak, G. C., Pierson, C. T., & Terenzini, P. (2003). Experiences and outcomes of first-generation students in community colleges. *Journal of College Student Development, 44*(3), 420–429.

Peteet, B., Montgomery, L., & Weekes, J. C. (2015). Predictors of imposter phenomenon among talented ethnic minority undergraduate students. *Journal of Negro Education, 84*(2), 175–186.

Pomerantz, E. M., & Grolnick, W. S. (2017). The role of parenting in children's motivation and competence: What underlies facilitative parenting? In A. Elliot, C. Dweck, & D. S. Yeager (Eds.), *Handbook of competence and motivation: Theory and application* (pp. 566–585). New York, NY: Guiliford Press.

Priniski, S. J., Hecht, C. A., & Harackiewicz, J. M. (2018). Making learning personally meaningful: A new framework for relevance research. *Journal of Experimental Education, 86*(1), 1–19.

Rheinschmidt, M. L., & Mendoza-Denton, R. (2014). Social class and academic achievement in college: The interplay of rejection sensitivity and entity beliefs. *Journal of Personality and Social Psychology, 107*(1), 101–121.

Schwartz, S. E. O., Kanchewa, S. S., Rhodes, J. E., Cutler, E., & Cunningham, J. L. (2016). "I didn't know you could just ask:" Empowering underrepresented college-bound students to recruit academic and career mentors. *Children and Youth Services Review, 64*, 51–59.

Smallets, S., Townsend, S. S. M., & Stephens, N. M. (2018). *Uncovering the light and dark sides of grit by separately considering its underlying facets.* Manuscript submitted for publication.

Smith, A. (2018, June 26). Re-evaluating perceptions about first-generation college students and their academic engagement. *Inside Higher Ed.* Retrieved from https://www.insidehighered.com/news/2018/06/26/re-evaluating-perceptions-about-first-generation-college-students-and-their-academic

Stebleton, M. J., & Soria, K. M. (2013). Breaking down barriers: Academic obstacles of first-generation students at research universities. *The Learning Assistance Review, 17*(2), 7–19.

Stephens, N. M., Hamedani, M. G., & Destin, M. (2014). Closing the social-class achievement gap: A difference-education intervention improves first generation students' academic performance and all students' college transition. *Psychological Science, 25*(4), 943–953.

Stephens, N. M., Markus, H. R., Fryberg, S. A., Johnson, C. S., & Covarrubias, R. (2012). Unseen disadvantage: How American universities focus on independence undermines the academic performance of first-generation college students. *Journal of Personality and Social Psychology, 102*(6), 1178–1197.

Stump, G. S., Husman, J., & Corby, M. (2014). Engineering students' intelligence beliefs and learning. *Journal of Engineering Education, 103*(3), 369–387.

Tibbetts, Y., Harackiewicz, J. M., Priniski, S. J., & Canning, E. A. (2016). Broadening participation in the life sciences with social–psychological interventions. *CBE-Life Sciences Education, 15*(3). doi:10.1187/cbe.16-01-0001

Torres, V., Reiser, A., LePeau, L., Davis, L., & Ruder, J. (2006). A model of first-generation Latino/a college students' approach to seeking academic information. *NACADA Journal, 26*(2), 65–70.

Tough, P. (2014, May 18). Am I supposed to be here? Am I good enough? *The New York Times Magazine*, p. 26.

Trevino, N. N., & DeFreitas, S. C. (2014). The relationship between intrinsic motivation and academic achievement for first generation Latino college students. *Social Psychology of Education, 17*, 293–306.

Walton, G. M., & Brady, S. T. (2017). The many questions of belonging. In A. Elliot, C. Dweck, & D. S. Yeager (Eds.), *Handbook of competence and motivation: Theory and application* (pp. 272–293). New York, NY: Guiliford Press.

Walton, G. M., & Cohen, G.L. (2007). A question of belonging: Race, social fit, and achievement. *Journal of Personality and Social Psychology, 92*(1), 82–96.

Walton, G. M., & Cohen, G. L. (2011). A brief social-belonging intervention improves academic and health outcomes of minority students. *Science, 331*(6023), 1447–1451.

Weiner, B. (1985). An attributional theory of achievement motivation and emotion. *Psychological Review, 92*(4), 548–573.

Winograd, G., & Rust, J. P. (2014). Stigma, awareness of support services, and academic help-seeking among historically underrepresented first-year college students. *The Learning Assistance Review, 19*(2), 17–41.

Yeager, D. S., & Dweck, C. (2012). Mindsets that promote resilience: When students believe that personal characteristics can be developed. *Educational Psychologist, 47*(4), 302–314.

Yeager, D. S., Walton, G. M., Brady, S. T., Akcinar, E. N., Paunesku, D., Keane, L.,...Dweck, C. (2016). Teaching a lay theory before college narrows achievement gaps at scale. *Proceedings of the National Academy of Sciences*, 1–8.

Zimmerman, B. (1990). Self-regulated learning and academic achievement: An overview. *Educational Psychologist, 25*(1), 3–17.

Zimmerman, B., & Schunk, D. H. (2008). Motivation: An essential dimension of self-regulated learning. In D. H. Schunk & B. Zimmerman (Eds.), *Motivation and self-regulated learning: Theory, research, and applications* (pp. 1–30). Mahwah, NJ: Erlbaum.

8

Mentoring and Forming Relationships With Students

What Is a Mentor? How Is a Mentor Different From an Advisor?

A mentor is similar to a coach, except that mentors are not sports coaches or life coaches. Rather, mentors are academic coaches. They are there to mentor, guide, and advise students as students make their way through their academic journeys.

Mentors should indeed do some of same things that academic advisors do, such as discussing and planning coursework with students, deliberating about careers, assisting with career planning, and discussing the academic steps that students should be planning for and carrying out in order to plan for particular careers.

But mentors need to do more.

Mentors need to be on the lookout for and to remind students of the steps that students do not know to think about and to plan for. For example,

Teaching STEM to First Generation College Students, pages 93–104
Copyright © 2019 by Information Age Publishing

mentors need to educate and remind students of the importance of networking, research, and internship opportunities. Mentors need to educate and remind students about the importance of forming relationships with faculty and of the need to solicit letters of recommendation from faculty.

To me, this characteristic of mentoring is really all about being an effective guide or navigator, to help students navigate the unfamiliar and sometimes rocky terrain of higher education, to make sure that students not only complete college, but that they get properly situated into the appropriate graduate school, professional program, or career path.

Why Be A Mentor?

Being a mentor to first-generation college students is highly rewarding. It is rewarding to help students who have had fewer advantages along the way make it and succeed in STEM. Helping first-generation students means you will be helping level the playing field for low-income students, nontraditional college age students, and underrepresented minority students, all of whom are disproportionately first-generation college students (Christensen, 2016; Larose, Chaloux, Monaghan, & Tarabulsy, 2010; Schwartz, Kanchewar, Rhodes, Cutler, & Cunningham, 2016).

But, more than just knowing you are helping deserving students and helping to diversify STEM, it can be personally rewarding mentor first-generation students. Watching students grow, develop, evolve, and overcome obstacles is highly rewarding. But nothing can compare to the emotional high of helping first-generation students realize their potentials and attain their heartfelt desires and goals.

Mentoring Students Who Are Struggling

Confession. Most of this chapter is devoted to discussion of the mentoring of first-generation students who are *struggling* or who are *facing challenges*. Some of my comments are applicable to those of you who are working with students who are not struggling or facing challenges, but I write most of my comments for those of you who work with students who are facing obstacles and who really *need you* as their mentor.

Why? Oftentimes, first-generation college students get into academic troubles because of their lack of familiarity with the higher education landscape, their lack of study skills, and their often times reluctance to seek help. I am therefore devoting a significant section of this chapter to the topic of how to mentor students who are struggling. By this I mean, how to guide or coach students who are struggling *overall* in their STEM coursework. If however, you have a student who is struggling in one specific STEM course (e.g., one that you are teaching), I would suggest you read some of the suggestions I have provided in Chapter 5 which are more along the lines of how to advise first-generation students taking classes with you.

A Glass Half Empty is Also a Glass Half Full

Most folks reading this book have likely heard the expression, "A glass half empty is also a glass half full." I can't remember exactly when I first heard this expression, but I can vividly recall when I began to think about how helpful this expression was for me personally in terms of framing how to think about and approach working with my undergraduate students.

All of us who work with college age students are likely accustomed to observing young people who are still growing and developing. We observe them with their strengths and weaknesses, with their faults and flaws, and with their youth and still developing talents. At times, we may envy them for the breadth of possibilities that lie ahead of them. However at other times, we may feel frustration at the mistakes that they make, especially at ones we can predict and foresee.

Long ago, I realized that I could dwell on the mistakes, faults, and flaws of my students and focus on what was lacking in their academic "glasses," or I could instead focus on the talents, abilities, and skills that they brought with them, on how we could work together to fill up their figurative or academic "glasses." I could look at students' transcripts, with their flaws and mistakes (poor grades), and tell students they had no shot of achieving their goals, or I could instead talk about all the hard and painful work they would need to do in order to move forward to work towards their goals (Yelland, 2015).

So I decided back then, that I would focus on the "half full" glasses. Without ignoring or sugarcoating what a student had already done, I would focus on what the student needed to do to move forward in order to achieve success and in order to explore and accomplish his or her goals (Rodriguez, 2015).

First-Generation Strengths and Weaknesses

As readers of this book, you would probably not be surprised to learn that one of the first topics I talk to students about (in terms of how to change and improve) is their *study skills*. I try to get students to understand that it is their first-generation college background that has likely led them to come to college lacking effective study strategies and that one way for them to work towards filling up their academic "glasses" is to begin to work towards becoming better self-regulated learners so that they can start to be more successful in their STEM classes.

At the same time, I also talk to first-generation students about resilience. Their resilience is their asset and strength. It is what has kept them going during economic challenges, family crises, and academic challenges. However, their never give-up attitude may also have led them to stick it out in a course when they should have withdrawn. Or it may have led them to repeat the same ineffective study habits over and over when they lacked knowledge of a better way to study.

I try to help students understand that they are going to need their resilience and determination even more now in order to recover from prior mistakes and that their resilience and determination is something they are lucky to have acquired, as it is something that cannot be taught or learned in a classroom (Horowitz, 2017). I tell them that now is the time to use the experience, resilience, and inner strengths that their life journey has given them, but to use them more strategically.

I am not trying to convince you that you should use my exact arguments when you work with and mentor students. Rather, what is important is that your students are able to see their background, socioeconomic status, parents' education, work experience, traits, habits, and so on, as potential sources of strength, as glasses that are "half full," rather than "half empty" (Dittmann & Stephens, 2017). This reframing is important because first-generation students who need mentoring sometimes have negative messages sent their way, telling them they are not good enough or that they are not going to make it (Gabriel, 2018). It is important to look clearly and honestly at where a student is at academically (Rodriguez, 2015) and to accurately assess what the student's chances are and what he or she can and cannot accomplish. But, how the situation is framed (from a positive versus negative perspective) is crucial (Cohen, Steele, & Ross, 1999, Dittmann & Stephens, 2017; Packard, 2015).

Lead With Your Intentions (Their Success)

When I begin a conversation with a student knowing that we are about to review and assess his or her transcript (which has poor grades on it), I make sure at the outset that my tone of voice and my language accurately convey what my intentions are. My intentions are always the *success of the student.* Starting with these intentions, I make it clear to the student that I am not there to judge, attack, or make him or her feel badly, that any talk about strengths and weaknesses is solely for the purpose of coming up with solutions and practical strategies to *help the student succeed.* Essentially, I want to start off the conversation making sure the student knows that I am on his or her side.

What do I mean by success, that I want the student to succeed? To me, success does not necessarily mean that the student gets into Harvard Medical School. Success to me means that he or she will complete a bachelor's degree and will have put in place a practical plan towards entering a career path where the student can find satisfaction and fulfillment. By this I mean I want my students to experience economic security, and I want them to be emotionally and intellectually fulfilled.

Listen Without Judgment

I try to leave my own emotions outside of the room when I meet with students. Sometimes, students tell me about decisions they have made that turn out to be bad ones or choices they elected that I might not agree with. Or, they tell me about family and personal situations that have negatively impacted their academics. I try to remain as neutral as possible in these situations.

However, I don't see my attempts to remain neutral and nonjudgmental as comparable to the role of a therapist. Rather, I see my efforts as a practical sort of neutrality, and I see myself as sort of a structural engineer. My job is to listen and take in the facts of the student's situation from as practical a standpoint as I can, without judgment, so that I can, together with help from the student, accurately assess what is the best course of action for the student.

The Science Biography of the Student

I typically start conversations with students by taking their "academic histories." I have students begin with descriptions of their high school science and math backgrounds, before moving forward to find out what they are majoring in, what courses they have enjoyed, and what courses have been roadblocks for them.

I also ask about other aspects of their lives that might have impacted their journeys, like how many hours per week they work at a paid job, what types of family responsibilities they have, and what particular issues may have been going on in their lives at different times when their GPA may have taken a dip.

I also try to find out what students are passionate about, what they wish for as an ideal, or even fantasy career goal, and what their current career goal is that they are striving towards.

Providing Honest Feedback

After (quickly) learning as much as I can about a student's background and STEM journey, I try to honestly convey to the student what I think are realistic goals. For example I might tell a student, given your current GPA, you are not going to get into a medical school in the United States, but I think you could get yourself into a place where you would be competitive for osteopathic medical schools (D.O. schools). Let's talk about what you will need to do (in terms of coursework and MCAT preparation) to make yourself competitive for D.O. schools. Or I might say, while it is true that your overall GPA meets the stated requirements of this specific nursing program, that particular program is very competitive. Let's talk about other, similar programs which also have good reputations, but for which I think you will have a better shot of getting in.

Practically Speaking

When I get to the planning stage of the conversation with students, I think of the conversation as consisting of three parts: assessing the present, planning short term goals, and planning long term goals.

Assessing the Present

To assess the student's present state of affairs, together with the student I attempt to summarize or synthesize the strengths and weaknesses of the student's academic profile. For example, a student may have a strong transcript, but may lack coursework in a specific area, or may lack required job, intern or volunteer experiences. A student may have a few poor grades scattered across his or her transcript, while another may have poor grades early in his or her academic career, and a third may have one particularly abysmal looking semester.

Planning Short-Term Goals

After assessing the student's present, I then discuss short-term plans with the student, what he or she should plan to do in the next semester or two to work towards longer term or bigger picture goals. Maybe he or she needs to repeat a course. Maybe he or she needs to audit or sit in on a prerequisite course (that he or she is not very strong in) in order to prepare to take a future, more advanced course. Maybe the student needs to plan out which requirements he or she is going to do when, and how to space them out, so that they will be completed most successfully.

Frequently, I observe students who move directly from prerequisite courses to more advanced courses even when they have performed very poorly in the prerequisite courses. This is in part because most students fail to appreciate the foundational role that prerequisite courses play in setting them up for success or lack of success in more advanced courses. However, I often find that when students move on to more advanced courses when they are not prepared for them, they do so simply for financial reasons, because they don't have the money to repeat prerequisite courses. Because financial aid only pays for students to repeat courses when they fail them, low-income students (who are disproportionately first-generation students) who are weak, but passing, in prerequisite material must move on to more advanced courses and end up doing so at tremendous disadvantage.

More importantly, however, when first-generation students complete prerequisites and move forward poorly prepared, they generally have no idea to what degree their lack of prerequisite knowledge is going to impact them. While it is probably true that most college students don't appreciate the value and sequencing of prerequisites, many first-generation students that I meet seem entirely ignorant of it. I have met quite a number of first-generation students who are seemingly completely unaware of the existence of prerequisites and due to computer errors have signed up for courses for which they have not met the prerequisites.

It is important therefore to share with first-generation students the truth regarding how much their prior, poor performance in a prerequisite course can impact their future performance, that they should not rush forward into courses they are not ready for, as this only causes further damage to their GPA's and transcripts. Rather, they should go back and repeat prerequisite courses if they can. And when they can't repeat courses, they should audit, or take off time, and review prerequisite material on their own. Rushing, in fact, only slows things down for them in the long run.

Formulating Long Term Goals

I find that some students are in a hurry and have the (crazy) idea that they just want to graduate, at all costs. They are burnt out and tired of school. If they only have a few required courses left, they only see a short-term goal of getting a degree. They then just make an unrealistic plan to pass all their courses at any cost, and don't even think about trying to do well in them. They just want to get out.

When I meet students who are just looking to get their degrees and get out, I try to remind them that they are not likely to be much better off with just their degrees. Rather, they should slow down and try to do things the "right way." They should try to get good grades and try to finish more slowly. They need to ask themselves where they are going to be when they have that bachelor's degree in hand. They will likely be in the same place they are now with the same dead end job.

Sometimes I have even had students tell me, that they just want to pass their last couple of science classes, whether they learn the material or not, and that then, later on, when they study for the MCAT, they will worry about learning the material!

When I hear this, I think to myself, what! Are you crazy?

But out loud, I calmly say, it doesn't work that way. You need to start learning the material now. The MCAT is about 10 courses worth of material. You cannot learn it all from scratch, while at the same time reviewing and preparing for the MCAT.

Students need to learn to slow down and start thinking long term. If the long-term goal is to go to graduate school, it is while the student is still working on his or her undergraduate major that the student needs to start planning out a step-by-step strategy. If the short-term goal is the one or two STEM courses the student is taking each semester (with the goal of pacing them out, auditing when necessary, and getting good grades in them), then here are some questions you and the student can explore together to help the student plan towards his or her *longer term* goals:

1. What bachelor's degree (what major) are you working towards?
2. When you graduate, what will your science GPA (or the GPA your desired graduate program cares about) look like?
3. When you get your bachelor's degree, will you be ready to apply to graduate programs? Will your GPA be competitive? Are there specific, additional prerequisites that you will need? Are you going to need to repeat some prerequisites? Are there some extra STEM courses you can take to make yourself a stronger candidate both in terms of content and in terms of GPA?
4. Do you need to consider a post-baccalaureate program? Will you be competitive for post-baccalaureate programs? Are there some extra STEM courses you could take to make yourself a stronger candidate both in terms of content and GPA?
5. If you are going to need to take additional courses, do a post-baccalaureate program, or begin to study for a standardized exam (like the MCAT), do you have a plan of how you are going to sup-

port yourself financially during this time period? How long do you expect this period of study to last?

Supporting Students During Their Times of Struggle

When working as a mentor of first-generation students, you will find that these students face struggles that are often different from those of their continuing generation classmates. Some first-generation students may be immigrants, low-income, live at home, have family responsibilities, have full-time jobs, and/or have families of their own, just to name some of the challenges they may face (Christensen, 2016; Cox, 2009).

Sometimes in order to effectively offer academic and career advice to students, it can be helpful to be informed about the significant nonacademic responsibilities that are in their lives, such as job and family responsibilities (Chan, 2010). However, you may not feel comfortable or properly trained to deal with some of the issues that students bring up that impact their academics. For example, students may share sensitive issues with you related to their financial situations (e.g., homelessness, joblessness, possible eviction) and/or medical and legal issues that they may be dealing with (e.g., depression, anxiety, abuse, drug use, immigration status, etc.; Christensen, 2016) Unfortunately, I find that these types of issues are sometimes present in my students' lives and that they do come up when we attempt to speak honestly about why students are struggling in their coursework or about what challenges are preventing them from having more organized, effective, and successful academic work schedules.

You should never feel responsible or obligated to offer counsel to students who are dealing with issues that you have not been trained to handle. It is best in these situations to encourage students to seek appropriate medical, mental health, or legal help. (In Chapter 5, I write about how to do this gently and carefully.) I would encourage you to keep information about these types of resources handy and ready, so that you can calmly and gently refer students to the proper resources should the need arise. However, it is also important to respond to students stories appropriately, by taking these stories seriously (not questioning their veracity), by treating the student in question with respect, and by responding without judgment.

When these situations come up, I do not try to be a therapist, but rather I try to play the role of *empathic* engineer. By this I mean I simply try to be a sympathetic, but practical, problem solver. I acknowledge the situation and the awfulness of it, and suggest resources for support and help. But I also try to treat the situation in a neutral way, and try to talk about strategies, specifically academic strategies. For example I might say, given your family situation in which you cannot study at home, how could you plan a place where you can study in quiet for long periods of time? Or given your financial situation, what kind of job could you get where you could do something that would not just pay the bills, but that would also help your resume or graduate school application (e.g., could you get paid to work in so and so's lab, work as a teacher's assistant, or do some chemistry tutoring)?

Praise and Positive Feedback are Important

As faculty, we are trained to seek out, mark, and correct students' errors. It is easy to forget to do the opposite, to pay attention to, notice, and praise students' accomplishments and successes. I think it is important to fulfill both roles. It is important to praise students and show enthusiasm for their accomplishments, but also to show disappointment when they hit bumps in the road or fail. A balanced, realistic approach demonstrates to students a genuine caring, that we are truly invested in them, when things go well and when they don't.

One way to praise students that perhaps gets overlooked in STEM environments is to praise traits and skills that we associate with good research habits. For example, when a student shows curiosity, asks questions, or makes good observations, this is a great opportunity to praise the student's behaviors. These behaviors are the ones we want to encourage, but they are often behaviors that get overlooked as they are not typically actions that we can grade students on.

Staying in Touch and Going the Distance

At the beginning of this chapter, I wrote about what a mentor could and should be. In my mind, because the job of the mentor is to help the student navigate the unfamiliar world of higher education, the job of the mentor should not end when the student graduates college or even when the student gets into professional or graduate school (Erdem & Eytemur, 2008). As a mentor of first-generation students, I try to stay in touch with some students (those that I have been close to), even if only to check in once a year, and even if only to check in during their first couple of years of their graduate programs.

I do this, in part, because I know that the transition period when students first enter graduate school and the earliest years of graduate school can be difficult. Having someone who knows them well and who knows their prior academic journeys can be helpful during those early years. But,

I also keep in touch because I know that when first-generation students get to graduate and professional schools, they will return to being first-generation students all over again. They will be starting graduate programs in which many of their peers (Gardner & Holley, 2011; Kniffin, 2007) have family members with graduate degrees. They will once again be in new and unfamiliar habitats where, in addition to the adjustments that all their classmates are going through, they will have to once again adapt to a landscape of unspoken rules. They may again go through the imposter feelings (Kniffin, 2007) that they experienced as undergraduates.

References

Chan, A. (2010). *Inspire, empower, connect: Reaching across cultural differences to make a real differences.* New York, NY: Rowan and Littlefield.

Christensen, W. M. (2016, September 9). Advising as activism. *Inside Higher Ed.* Retrieved from https://www.insidehighered.com/advice/2016/09/09/importance-advising-first-generation-students-essay

Cohen, G. L., Steele, C., & Ross, L. D. (1999). The mentor's dilemma: Providing critical feedback across the racial divide. *Personality and Social Psychology Bulletin, 25*(10), 1302–1318.

Cox, R. D. (2009). *The college fear factor: How students and professors misunderstand one another.* Cambridge, MA: Harvard University Press.

Dittmann, A., & Stephens, N. M. (2017). Interventions aimed at closing the social class achievement gap: hanging individuals, structures, and construals. *Current opinion in psychology, 18,* 111–116.

Erdem, F., & Aytemur, J. O. (2008). Mentoring: A relationship based on trust: Qualitative research. *Public Personnel Management, 37*(1), 55–65.

Gabriel, K. F. (2018). *Creating the path to success in the classroom: Teaching to close the graduation gap for minority, first-generation, and academically unprepared students.* Sterling, VA: Stylus.

Gardner, S. K., & Holley, K. A. (2011). "Those invisible barriers are real:" The progression of first-generation students through doctoral education. *Equity and Excellence in Education, 44*(1), 77–92.

Horowitz, G. (2017). First generation college students: How to recognize them and be their ally and advocate. *Journal of College Science Teaching. 46(6),* 6–7.

Kniffin, K. M. (2007). Accessibility to the PhD and professoriate for first-generation college graduates: Review and implications for students, faculty, and campus policies. *American Academic, 3,* 49–79.

Larose, S., Chaloux, N., Monaghan, D., & Tarabulsy, G. M. (2010). Working alliance as a moderator of the impact of mentoring relationships among academically at-risk students. *Journal of Applied Social Psychology, 40*(10), 2656–2686.

Packard, B. W.-L. (2015). *Successful STEM mentoring initiatives for underrepresented students*. Sterling, VA: Stylus.

Rodriguez, E. (2015, May 18). Mentors matter: Individual achievement goes only so far. *The Chronicle of Higher Education*. Retrieved from https://www.chronicle.com/article/Mentors-Matter/230147

Schwartz, S. E. O., Kanchewa, S. S., Rhodes, J. E., Cutler, E., & Cunningham, J. L. (2016). "I didn't know you could just ask:" Empowering underrepresented college-bound students to recruit academic and career mentors. *Children and Youth Services Review, 64*, 51–59.

Yelland, H. (2015). Discovering assets. In J. Fletcher, A. Najarro, & H. Yelland (Eds.), *Fostering habits of mind in today's students: A new approach to developmental education* (pp. 17–52). Sterling, VA: Stylus.

9

Advice for Undergraduates

Am I Just Not Smart Enough?

Chances are if you are an undergraduate reading this chapter, you may be struggling to succeed in your college level science and math courses. You may feel overwhelmed and may be wondering what you are doing wrong. You may look around at other students who seem confident, happy, and successful and wonder what makes you so different? Maybe you are not smart enough? Maybe you are not cut out for science? Maybe you should give up and choose a different major? Or give up entirely on your education and just focus on finding a better paying job?

Well, my answer to you is "*No!*" You are not stupid. You are not more stupid or smarter than those around you! You are not lazier or less of a good science student. In fact you are likely more motivated and hardworking than others! Chances are if you are working hard but struggling, and feeling like you are struggling more than other students around you, there are one or two possible reasons why:

1. You may be in classes with students who came out of high school better prepared than you (in terms of science and math). You

Teaching STEM to First Generation College Students, pages 105–124
Copyright © 2019 by Information Age Publishing
105

may have been a very strong student in high school. You may have worked harder than many of your high school classmates and you may have done very well in your high school science and math courses. But—some of your college classmates likely had more competitive science and math courses than you and/or more advanced science and math courses (including AP courses).

See below for more info about what you can do to help yourself in this situation.

2. If you are a first-generation college student (the first-generation in your family to complete a bachelor's degree) or the first in your family to complete a bachelor's degree in science, technology, engineering, or mathematics (STEM), you may find yourself at a disadvantage relative to some your classmates who have family members at home who have college degrees in STEM. Why? Simply put, you have a lack of savvy or experience with college and the process it takes to be successful in college level science. (I will explain more about this below.)

The good news is that learning the "tricks of the trade" of how to be a successful STEM student is not difficult! The bulk of this chapter is all about teaching you how to acquire these tricks—the savvy or skills that other students possess that enable them to be successful.

This savvy or skill set has a technical name—"self-regulated learner." You need to learn how to become a self-regulated learner.

But before I explain what a self-regulated learner is, I want to briefly talk about what to do if you high school did not prepare you adequately for college level STEM courses.

Gaps in Preparation

First of all, know that you are not alone.

If you are a transfer student, you may have taken certain prerequisite math and science courses at a community college, have just transferred to a 4-year college, and are now experiencing a rude awakening; you got credit for prerequisites and are supposedly prepared for the classes you are in, but you find yourself in over your head.

Or if you went straight from high school to a 4-year college, you went from a situation in which you were a strong science/math student, but now are taking classes with students who went to top (magnet type) high schools where they had AP science courses and other advanced science and math electives.

Unfortunately, this is a common problem that many STEM students face, that when they begin at a 4-year college, they find that there is a gap between their background preparation and the new courses they enroll in. And this problem is especially common for students who come from urban environments and/or low income communities.

Don't let this get you down! Know, that your background is your strength, something that can really benefit you! Your background likely taught you to work hard, to be persistent, and to never give up. This trait, termed *resilience*, is something you are really going to need and benefit from, now and in the future. Your persistence and resilience can give you, a first-generation college student, an edge over other, perhaps better prepared, STEM students.

This is the good news.

But the bad news is that there are no shortcuts to getting yourself caught up. You need to work harder than other students in your introductory science and math courses. If your university offers slower paced or more remedial type courses in science and math, I strongly suggest you take them (if you can afford to do so). A solid foundation in math (particularly in arithmetic and especially in algebra) is going to really matter if you want to take courses like calculus, but also if you want to take courses like general chemistry and introductory physics. (If you cannot afford to enroll in slower paced courses, try to find courses you can sit in on for free, like large lecture courses where one extra student doesn't matter. Or look for online lessons that are free.)

Becoming a Self-Regulated Learner

But what is a self-regulated learner, why is being one important for your success, and how can you become one?

A self-regulated learner literally is a person who regulates his or her own learning. What does that mean? That means being "in touch with yourself." Yes, this does sound a little touchy feely or strange, but what do I really mean? I mean you need to have a running conversation with yourself in your head, where you check in with yourself multiple times each day and ask yourself where you are at in terms of your learning and studying. Are you following what is going on in class? Which topics are you following and which are you not? What actions do you need to take to be more effective in your study habits? Are there instances in which you need to modify or change your behaviors in order to be most successful?

Here are some specific STEM questions you should be asking yourself along with matching, specific actions you should take if your answer to a particular question is yes.

▬▬▬▬

Mastering the Course Content

- ▪ *Do you find that reading your textbook takes way too much time and/ or that you learn and remember very little of what you have read?* Part of what happens when we read a text is that we tend to zone out without realizing it even if our eyes keep reading (McGuire & McGuire, 2015). The best way to combat this to try to read in as *active* a way as possible.

    Begin with a *goal* and with a pencil in hand! Look at the chapter and section headings to see what the topic is and read with that topic in mind. As you read, make brief notes and/or summaries of what you are reading (in the margin or on a separate piece of paper). Trying *not* to copy down every detail of what you read, but see if you can capture the idea of each paragraph with no more than one sentence. If what you are reading is very visual or conceptual, it may be better to try to capture it with a diagram or picture rather than with a sentence.

    But either way, do *not* use a highlighter. A highlighter can encourage you to passively mark text, whereas a pencil will force you to synthesize what you read and actively write it out in your own words. Similarly, it is better to physically hold a pencil in your hand and write or draw, rather than type on the computer (McGuire & McGuire, 2015). (For more ideas about how to read actively check out Chapter 5 of an excellent book entitled *Teach Students How to Learn* by McGuire and McGuire.)

- ▪ *Do you find yourself lost during lecture?* Try skimming the book for 5 minutes before class each day so you can walk into class with some idea of what to expect. Then review your notes immediately after class, quickly entering missing information and questions into your notes as soon as possible after class.

- ▪ *Are you unable to take good notes during lecture?* Consider tape recording the class so that you can take time during class to focus on listening and then take detailed notes later. If your instructor uses PowerPoint during lectures and doesn't provide advance copies of the slides, bring a copy of your textbook to class so you can write or draw quickly *in pencil* directly in your textbook without having to copy down all the sketches or diagrams that are in the PowerPoint, but that are also displayed in your textbook.

At the minimum, having a copy of your textbook will allow you to jot down the figure numbers that your instructor is focusing on so that when you review your lecture notes you will know which figures to be looking at. (If you only have an electronic copy of your textbook, bring it to class on a tablet or laptop so that you can write in the e-book similarly to how you would write in a hard copy of the textbook.)

▪ *Are you feeling confused about a* specific *topic?* Your impulse might be to first look on the internet to get help (as the internet is available 24/7 and is completely anonymous). But often if you have a specific area of confusion, searching the internet can waste a tremendous amount of time and can end up becoming like looking for a needle in a haystack where you have to sift through tons of useless information. However when you need clarification of a specific topic, it is often faster and more effective to ask a human, like an instructor, tutor, or even classmate for help because they can likely quickly hone in on the specific area you need help with.

▪ *Do you need an* overview *or* review *of a topic?* An online video can be very helpful in this instance. Two very good websites that have short videos on lots of specific, college level STEM topics are freelanceteacher.com and khanacademy.org. Check them out.

▪ *Do you have a basic idea of a topic, but don't feel like it is sticking with you?* It sounds like you need *rehearsal* of the topic. Maybe doing some practice problems from the back of your textbook chapter is the right way to go depending on what the subject is. Another option is to write out an explanation in the form of an essay, a diagram, or a concept map.

▪ Maybe explaining the topic out loud to a classmate is most effective depending on the topic. (Pick a classmate who will push you to make sure you explain yourself thoroughly.) Another option is just to explain the concept out loud to yourself (in an empty classroom, a private bedroom, or a bathroom). Explaining out loud has an added advantage beyond just helping you rehearse and learn a topic. It will also help you notice and become aware of what gaps you may have in your understanding (McGuire & McGuire, 2015). Keep in mind that explaining in writing is basically what you will be asked to do when you take an exam (you explain something to the instructor; B. Coppola, personal communication, 2016). So a dress rehearsal or practice at explaining is way better than walking in cold to an exam never having practiced your "explanation" ahead of time.

▪ *Are you feeling lost in a* broader *or more way?* A tutor is the best solution in this case. Your institution hopefully has free or low cost tutoring available through a learning or academic success center. If you don't know what your campus has available, now is the time to find out!

▪ *Do you feel like you don't have a clue about* how to approach studying *for a particular course?*
A number of suggestions:
- Get to know classmates who seem like they know what they are doing and try to mimic them.
- Find students who have successfully completed the class and ask them for pointers.
- Join a student club that supports science majors, especially women and/or minority science students. Clubs are often a good source of insider information about what it takes to be successful in a course.
- Consult your instructor and his or her course syllabus. Yes, your instructor may be scary, but he or she probably knows a lot about why students struggle in the course and what the best practices for success in the course are.

▪ Find it hard to get help on your campus? Tutors and instructors seem too busy?
Save their time and yours by clearly marking in advance what and where your questions are. Do this by marking up your class notes, textbook and homework problems to indicate the places you are getting stuck. Ways to mark things up can be putting by question marks in the margin in pen, by highlighting specific text or by using colored Post-Its. These methods will allow you to quickly sort through and find where you are having issues to avoid having to fumble or rush when sitting in an instructor's office.

Another option you can try is to take advantage of downtime during lab. Often experiments have waiting periods. This is a good time to try to connect with your classmates, share information, and get help from them. It may also be true that during waiting periods, your lab instructor may be receptive to you asking questions.

Problem Solving

▪ Do you find that you can do basic problems from within the textbook chapter, but stumble when you attempt the end of chapter problems?

- A study group can be a good idea here where different students can provide different inputs and perspectives and help each other get through sticking points.
- Another approach can be to look back to the chapter material itself each time you find yourself hitting an end of chapter problem that you get stuck on. Try to find and review the relevant text material that addresses the same topic that the end of chapter problem focuses on.
- Alternatively try to find an *in* chapter exercise (usually a much easier example, a sort of warm up question) on the same topic. And do the practice problem first as a practice to remind yourself how to approach the *end of* chapter problem.
- If none of the above strategies work for you, then you are likely in over your head and you need some help from an instructor or TA. You need to go to office hours and if that doesn't provide the help you need, you need some one on one help in the form of tutoring.

■ Are you able to do end of chapter problems, but still struggle during exams?

- Keep in mind that when you are solving problems (for example your assigned homework problems from the back of the chapter) the goal of your instructor was most likely NOT that you simply determine the correct answers to specific questions. Rather his or her goal was most likely that you learn about the processes by which you solve each problem and/or the principles that underlie each problem (McGuire & MgGuire, 2015). So you need to ask yourself what are you really learning and accomplishing when you solve homework problems:
  • Do you mostly get the answers right, but are not sure why they are right? You need to be able to get to the point where you can explain your answers out loud (e.g., to another person) and do it with confidence.
  • Are you someone who uses the solutions manual as a crutch? By this I mean do you *overuse* the solutions manual? Do you look at it too soon before you really fully attempt to solve the each problem, so that you end up convincing yourself that you *can* solve the problems, but really you cannot solve them without having the solutions manual *open* in front of you? Hide the solutions manual and force yourself to fully attempt each problem and only use the solutions manual as a *check*, not as a cliff notes or study guide!

- Do you have a different issue where you *never* check your answers (never compare them to the solutions manual)? So you spend lots of time working and doing problem solving, but you never have any idea if the work you are doing is right or wrong? You may in fact be spending hours reinforcing the *wrong* ideas! I know you can't afford to buy a solutions manual, but I'll bet you can find it in your school library. If not, ask your professor to put it on reserve! You need to know that you are worth it and need this access in order to be successful STEM student.

– Have you practiced enough end of chapter problems? Don't just do the problems your instructor has assigned. Practice until you have done the majority of the most difficult questions in the end of the chapter. Practice until the difficult end of chapter problems feel really boring and until you can easily explain them to another student.

– Are you able to answer questions out of context? Make sure you practice problems out of context. Remember on exams, you won't know what chapter a particular question is from and this can catch you by surprise.

– Can you synthesize ideas from different chapters? Keep in mind that exam questions can be global or comprehensive in their content and ask you to pull together ideas from multiple chapters. Make sure you practice questions that pull content from different places and different chapters so you know that you can handle those types of questions.

– Are you exam ready? Try to get a hold of some sample exams written by your instructor so that you know what to expect format wise and content wise. But don't try to memorize specific questions from past exams. Instead, try to practice as many questions as you can similar in style and level of difficulty to those written by your instructor.

– Do you utilize your graded work as a source of feedback? If you are like many students that I know, you may avoid picking up or looking at graded work (especially if you have done poorly on an exam or quiz). But this is not what you should do if you are looking to learn about yourself and learn how to improve your knowledge and study habits. Reviewing an exam can be a valuable experience (Mcguire & Mcguire, 2015). You can see what topics you were weaker or stronger in. You can find out how and when you make careless mistakes and/or find out if you properly paced yourself timewise during an exam. Learning from their mistakes is something that the academically strongest students in your class are always doing—you need to be doing this too if you want to be able to reflect on your study habits and learn about which ones are working for you and which ones are not.

Every Course Is Different

Above, I described different strategies you can utilize to make sure you are most effective when you are engaged in different study behaviors such as reading, taking notes or problem solving. But another skill you need to develop is to figure out which study strategy to use when (Bauer, Kim, Zureick, & Lee, 2016). In other words, different study strategies are needed for different courses and for different topics within courses. Some courses are very content heavy. For example, you might take a biology course in which you need to learn a lot of new vocabulary or need to memorize all the bones of the human body. For learning vocabulary, *actively writing out* the terms and their definition on flash cards (not buying or photocopying flashcards) might be very useful. For memorizing the names of the bones, making many photocopies of a blank template of the human body and filling in the names over and over might be the way to go. Part of being a strategic studier or good self-regulated learner is to develop this skill of knowing *what to do when* in terms of your study strategies.

A Little Bit Every Day

As an instructor of organic chemistry, one thing I advise my students is to do a "little bit every day." I say this in part because it is of course better not to cram and to instead do a little bit of studying every day. But I say this to students for another reason. I find that all of my organic students (from strongest to weakest) often get stuck as they study and work through the material. Sometimes they get stuck when reading and working through conceptual ideas. But more often they get stuck when working on problem solving. But when they get stuck, they often get bogged down and cannot move forward. They then end up wasting a lot of time spinning their wheels (sometimes searching endlessly on the internet to try to solve their problems).

So therefore, I advise them to try to do a little bit each day so that they can check in each day with a human, either an instructor, TA, or tutor from our learning center. This can be in person or via email. Sometimes a quick question and quick targeted response can make all the difference. It can enable you to move forward and be much more efficient and effective in your studying.

This is a really important issue that first-generation college students need to consider. Many of you may be working part or full time and may plan to *consolidate* all of your studying into one or two days a week (e.g., for those days when you are not working or in not school). Try to think and plan ahead so that you do *not* schedule 10 hours of one day towards the study of one academic subject. Try as much as possible to do a little every day even if it is just for 10 or 15 minutes. Give yourself a few minutes each day just to figure out what you know and what you don't know for a given subject and to plan what you will do the next day to resolve the things you don't know. *If you get to the point in your studying where you can do this, you have become a master self-regulated learner!!*

Motivation, Anxiety, and Time Management

- Do you find that you sometimes are late to or miss class?
  - Perhaps you are very busy with the responsibilities of life, but it may also be that if a class is getting to you and making you feel bad about yourself, that coming late may be a coping mechanism for you—a way of avoiding acknowledging that you are struggling. Why? Because if you are late to class, it means you can spend less time in a setting or environment in which you feel scared, intimidated, depressed, confused, or so on. Furthermore, if you are late to a quiz or exam, it is then easy to justify to yourself as to why you didn't do well.
  - Ask yourself why you are not early or at least on time to every class. In order to become a successful student, you need to come to class on time or even early (Bauer et al., 2016)! The first remarks your instructor makes at the start of class are often the most important ones!
  - And if you are taking an exam—a fast-paced, very stressful, competitive style exam, you *owe* it to yourself to come early, to get the seat you want, to get settled, and so on so that you optimize your chances of doing your best.
- Do you panic or "blank out" during exams? Or even just run out of time?
  - One main reason why students panic or blank out is because they are not well prepared. Make sure you are over practiced! This cannot eliminate your anxiety, but it will help a good deal.
  - Rehearse the exam situation in advance:

Time yourself while practicing. Get a sample exam if you can or just some difficult problems. If working from a sample exam, pace yourself so your timing matches the points assigned to each question (e.g., a 10 point question on a 75 minute exam should be allotted 7–8 minutes only). You can also assign time lengths to non-exam problems if you do not have sample examples to practice from. Either way, when rehearsing, pace yourself very strictly so that when you face the real exam you don't let a very difficult question throw you or take over your time. If you can, rehearse in a classroom or even in the actual exam room itself. This will enable you to mentally prepare yourself for the setting so that it will be less intimidating on the day of the actual exam.

– Are you someone who has a history of anxiety during exams? Go see a health-care professional! (Your college likely has a medical facility that offers free health care to students.) A health-care professional may be able to prescribe medication that will help you and/or provide you with a diagnosis that will entitle you to special accommodations during exams. Alternatively, go to the disability office on your campus. Many psychological and medical conditions can entitle you to special accommodations during exams. Sometimes just being able to take your exam in a quiet room (away from other students) can be enough to enable you to feel calm and perform better.

Self-Regulated Learning

Let's talk again about self-regulated learning. Hopefully, you have just finished reading the various questions I have listed above along with my suggested strategies for addressing them. But remember, you need to be regularly and daily asking yourself those questions while you have conversations with yourself about your studying. Linda Nilson, author of a book entitled *Creating Self-Regulated Learners* (2013), describes self-regulated learning as the "deliberate practice" (p. 6) that leads to expertise. Imagine a professional athlete and all the years of training he or she needs to put in to become an expert. Along the way, the athlete must question each and every aspect of his or her performance and target specific areas that need tweaking, modification, and extra practice. Think of this as your job now, to assess everything you are doing and modify as needed. *That is what self-regulated learning is all about.*

Other Things to Keep in Mind

Don't Let Successful Students Scare You

Students who have college educated parents and/or who come from magnet high schools figure out how to adapt to college level science and math courses more quickly than first-generation college students. This is because

- these students may have had prior exposure to challenge (from high school) and therefore know how to respond more successfully to it, and
- these students may have parents who subtly taught them the tricks of the trade of how to do STEM in college.

Avoid students who you find unfriendly or intimidating, but see if you can find successful students whose behaviors you can emulate. If they are good self-regulated learners, try to copy and mimic their behaviors.

Don't Let Your Parents Get You Down

Many first-generation students tell me that their parents can sometimes be discouraging to talk to. Non-college educated parents (or those who didn't study science in college) often don't understand why their children are struggling to succeed in college science. These parents often tell their children that they need to work harder or that they are not working hard enough. But, if they see that their child or children are working hard and are still struggling, then they often tell the child or children something to the effect of, "Oh you must not be good enough to do science. Maybe you ought to do something else with your life."

I respectfully disagree with these types of comments from parents. If your parents didn't study science in college, then it is likely they have *no idea* of what it takes to be successful in college science. If they advise you to "work harder," my response is: "No! Work *smarter*, not *harder*!" Your parents may not realize that it is a certain savvy that one needs in order to be successful in college science. This "savvy" that I have described above (self-regulated learning) is not about working harder. But your performance as a first-generation student is also not about your aptitude or ability (if your parents try to tell you "maybe you are not cut out for this").

Chances are if you talk to your course instructors about how to be successful in their courses, they will have a lot of specific advice for you. They may even be experts in what one ought to do in order to be successful in their courses. So it is not disrespectful to your parents to set aside their well-meaning advice. Rather, it is more effective in this situation to seek the advice of *experts*, just like you might do if you needed surgery or if you needed to do an expensive repair of a car or computer. I hope you can show this chapter to your parents and explain to them that *my* expertise is all about how to be successful in college science, especially how to be successful as a first-generation college student.

Don't Be Afraid to Take Risks or Learn New Habits

As a first-generation college student, you may have grown up in a household where you were encouraged to fend for yourself and not rely on others. Traits like hard work, resiliency, and persistence may have been emphasized and fostered in you. These traits will serve you well when hard work and effort are demanded of you—and these traits are certainly important and necessary in order to be successful in STEM. But I have observed that successful students are also effective "help seekers." By this I mean, successful students don't think twice about visiting professors' offices, taking advantage of tutors, or speaking up in class when they have questions, and so on.

You may be thinking. I am shy. I hate asking for help. I hate even admitting I don't understand something or I need help.

I do empathize with you. I personally have a lot of trouble asking for directions and have been known to wander and waste lots of time, rather than admit I am lost. Asking for help finding something in a store is also something hard for me to do.

But what can I say, other than: "Get over it!" You are competing with people who are comfortable seeking help, specifically, academic help. There is much research out there that indicates that academic help seekers are more successful than others. (See Karabenick, 2003 and Szu et al., 2011 for examples from college level science classrooms.) In fact, academic help seeking is considered a form of self-regulated learning (Karabenick & Berger, 2013; Karabenick & Gonida, 2018; Newman, 1994). So push past your discomfort. Seek help: from classmates, TAs, instructors, tutors, whoever is useful to you. And try to remember that successful students don't have any qualms about asking questions, even stupid, annoying ones where it is evident and obvious to all that they are lost, not prepared, not listening, and so on. (Trust me, I have seen this many times.)

Maybe it would help for me to remind you that *you belong* where you are now (in college). You have gotten yourself there and you deserve to be there. You may be working part or full time at a job to help pay for school or to help support yourself while in school. And I'm sure that your hard earned money is important to you. So take advantage of any and all resources the school offers, even if it feels uncomfortable! Get your money's worth out of school. And if you find it difficult to find good sources of help or find that certain staff/faculty are not receptive, try asking other students where to go, and find who is most helpful in which types of situations.

Find a Mentor

One of the things that can be really helpful on your STEM journey is to find people to support you along the way. Finding supportive friends and family is important. (This can include classmates and students you meet through campus organizations and clubs.) But another potential source of support at your college can be faculty and other staff (like academic advisors and other administrators). The "grownups" at your college, if they are supportive and helpful, can be a great resource for you. Finding someone who has been through college already, who understands what higher education is all about, who maybe has him or herself majored in a STEM field and/or been a first-generation college student can be very helpful.

How to find such a mentor or ally? Start by paying attention to the professors who teach your classes. Some professors set a tone in class that encourages students to participate and ask questions (even if the class is large). On the other hand, some professors don't even respond to their email and are unavailable for office hours. Some professors have extensive resources online for students and/or encourage group discussion in class or online. Professors who send the message that they are open to students and that they are a partner in helping students succeed (Gasiewski, Eagan, Garcia, Hurtado, & Chang, 2012) are the ones you should seek out! Of course talking to your classmates about who have they found helpful is a good idea too. And it may very well be that your campus may have faculty who are part of a network of faculty who promote and advertise themselves as mentors and allies for first-generation college students.

Don't Assume College Is Just a Continuation of High School

It may have been the case that in high school you often were expected to memorize information and then simply spit it back on exams. College, however, is different:

▪ You will likely be expected to assimilate and absorb much more material at a much faster pace.

- You will be expected to do a good deal of your learning on your own outside of class. And as you may have figured out already from class or from what I have been saying above, your college professors are not going to give you much in the way of instructions about *how to study*. They are going to give you only an overview of the topics they expect you to know, but will then expect you to figure out on your own *how to study* and how to teach yourself the rest of the material (Bauer et la., 2016).
- Instructors will likely not provide review sheets or extensive in-class reviews.
- Exams will likely not be spit back and memorization. Exams will expect deeper, more sophisticated thinking on your part and may also expect you to integrate or synthesize together multiple topics and ideas. Exams may expect you to use what you know to solve problems more advanced than or unlike those you have seen in class or in the homework!

A 4-year college will also be very different from high school or community college in that it may be much larger:

- You may be in science classes with 500 other students. Professors will not know if you don't come to class, or if you are physically present, but not paying attention. And they are not going to check up on you to see if you are keeping up with the reading or doing your homework. It is your job now to make sure you *police yourself*—this is part of what it takes to be a self-regulated learner. And by the way, coming to class, coming on time and sitting in the front *will* give you an edge over others. Try to never miss class. If you cannot help but be late, show up anyway! This may be your only opportunity to hear the lecture. You miss crucial information by not coming to class.
- At your college, the staff to student ratio may be much smaller than you are accustomed to from high school—meaning there may not be as many people available to help you with finances, registration, advising, you name it.
- Faculty will be very busy. Their primary responsibility may be to their research, rather than to teaching. Faculty may be comfortable with students dropping in unexpectedly, but faculty may expect you to make an appointment or may only meet with students during fixed office hours. Faculty will generally not be available at the last minute, so it will always be a good idea to make an ap-

pointment in advance. During summers and winter intersession they may not be on campus at all. So plan ahead.

Don't Get Discouraged

Learning how to become a savvy or sophisticated science studier (a good self-regulated learner) is not something that happens overnight. It is a journey and a process that takes time. It is something that all STEM students need to do, not just first-generation college students. (Unfortunately "continuing" generation college students have a big head start.)

However in my experience working with lots of first-generation college students, this learning process can be stressful and painful. While you are in the process of figuring out "how to study," your grades may be suffering. You need to know that at the end of the journey you will come out stronger! And at the end of the journey you can be successful! Graduate programs will notice students who overcame obstacles and whose GPAs have an upward slope. Medical schools want students who have had challenges in their past and have been able to overcome them. (Students who have never had to think about how to study can flunk out when they hit their first year of medical school.) Hopefully your university advisors can talk to you about how to strategize and how to navigate so as to more effectively and successfully get through your journey and achieve your goals. (You can contact me about this as well.)

If you want to read more about how to survive and thrive as a science student, I would suggest you check out the excellent pointers written in Chapter 3 of an excellent book, *What Every Science Student Should Know*, written by former premed students (now med students) Bauer, Kim, Zureick, and Lee (2016). Another good resource for learning how to improve and become a better studier is a book called *Learn How to Study and SOAR to Success* by Kenneth A. Kiewra. In particular, I recommend Chapter 2 of this book which focuses on how to take notes in lecture and how to read a textbook effectively. However, other chapters focus on topics such as how to prepare for and learn from exams, how to deal with test anxiety, how to tackle courses that are very fact intense (like some biology courses).

Appendix
How to Approach Problem Solving the Right Way

Below is a handout that I include in my Organic Chemistry I syllabus to try to guide students as to how to approach problem solving. This handout summarizes the series of steps I advise you to take as you approach problem solving in any physical science or mathematics course. The handout also offers suggestions for how to assess yourself and your level of skill and how to obtain help when you need it.

The Drawing

At the top, left of the handout is a drawing or flowchart with boxes and arrows. This part of the handout illustrates the pathways I suggest you should take as you work on problem solving.

Start with the section of the drawing that is labeled with the Roman numeral I. Work on some introductory problems, then assess yourself and see if you

 a. need more practice at simple, introductory problems;
 b. need to back up and get help; or
 c. are ready to move on to more advanced practice.

When you are ready to do some advanced practice, you will move to Part II of the drawing. Again you will need to assess yourself to see if you

 a. need to keep practicing advanced problems;
 b. need to back up and get help (or maybe move back to simpler problems); or
 c. are ready to move on to the next topic, Part III of the drawing (where you start the next topic, but go back to Part I, and do the introductory problems for the next topic).

The Boxes

But how to know how to assess yourself or what to do when you find you are in trouble and need help? For these issues, I have given suggestions in boxes. These boxes are in the upper right and bottom parts of the handout, and are entitled "Ways to Assess Yourself" and "Help Seeking Guide," respectively.

Assessing Yourself

There are a number of good ways to see if you really know things as well as you think you do. One way can be to check an answer key, but a better

How to Approach Problem Solving

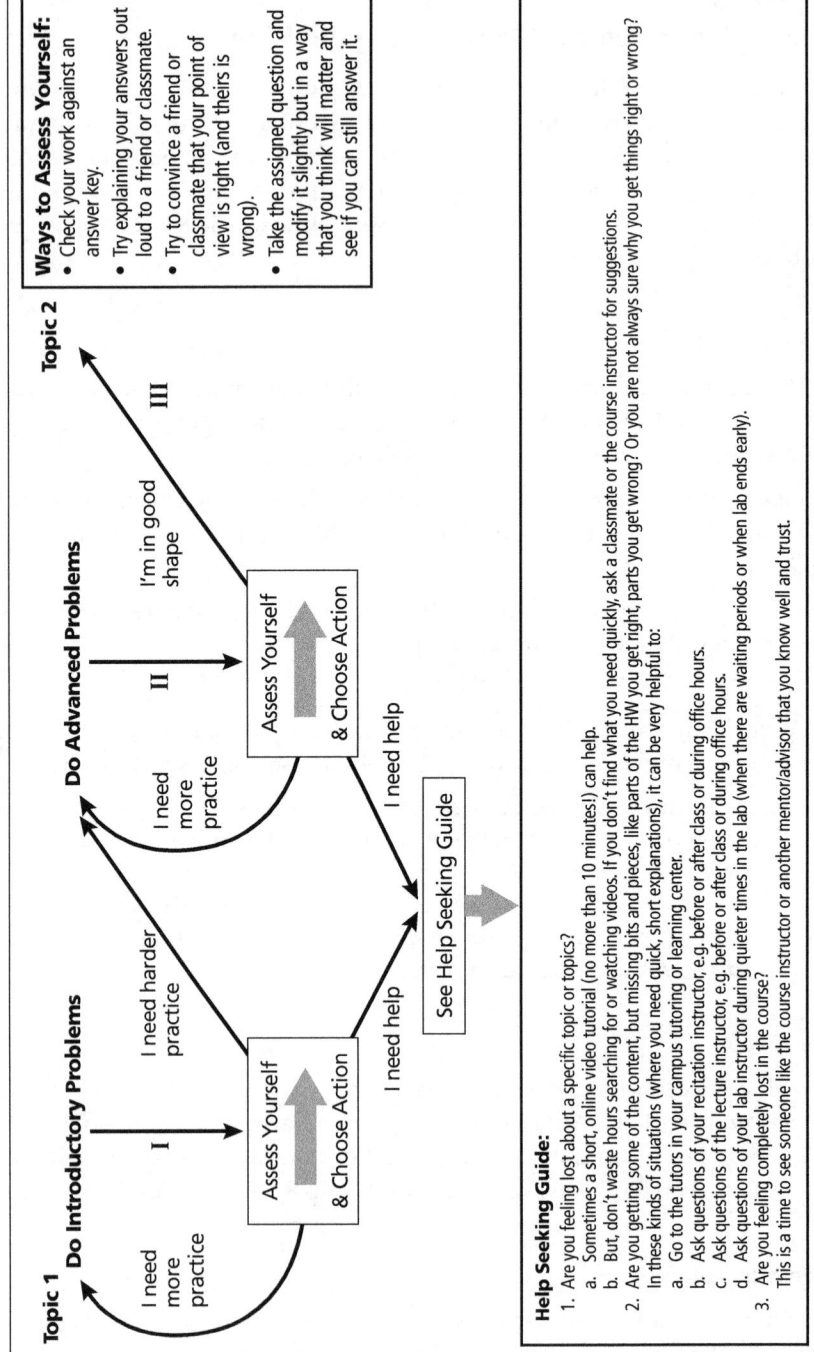

Ways to Assess Yourself:
- Check your work against an answer key.
- Try explaining your answers out loud to a friend or classmate.
- Try to convince a friend or classmate that your point of view is right (and theirs is wrong).
- Take the assigned question and modify it slightly but in a way that you think will matter and see if you can still answer it.

Topic 1
Do Introductory Problems

I
I need harder practice

I need more practice

Assess Yourself & Choose Action

I need help

Do Advanced Problems

II
I'm in good shape

III

Topic 2

I need more practice

Assess Yourself & Choose Action

I need help

See Help Seeking Guide

Help Seeking Guide:
1. Are you feeling lost about a specific topic or topics?
   a. Sometimes a short, online video tutorial (no more than 10 minutes!) can help.
   b. But, don't waste hours searching for or watching videos. If you don't find what you need quickly, ask a classmate or the course instructor for suggestions.
2. Are you getting some of the content, but missing bits and pieces, like parts of the HW you get right, parts you get wrong? Or you are not always sure why you get things right or wrong? In these kinds of situations (where you need quick, short explanations), it can be very helpful to:
   a. Go to the tutors in your campus tutoring or learning center.
   b. Ask questions of your recitation instructor, e.g. before or after class or during office hours.
   c. Ask questions of the lecture instructor, e.g. before or after class or during office hours.
   d. Ask questions of your lab instructor during quieter times in the lab (when there are waiting periods or when lab ends early).
3. Are you feeling completely lost in the course?
   This is a time to see someone like the course instructor or another mentor/advisor that you know well and trust.

way can be to test yourself orally by having to defend your argument to a friend. Even talking out loud to the wall works! I have spoken about this topic more extensively in the chapter above, but the handout below is a good outline or reminder to keep handy.

Getting Help

Seeking help is something trickier and again this is something I spoken about above. You need to get past your discomfort with admitting you need help, but you also need to think about how to be strategic. How are you going to maximize the use of the resources available to you on your campus (e.g., instructors, TAs, tutors, physical resources, etc.)? And be smart too about how you use online resources, which can be very useful, but can also be extremely wasteful of time as they don't target their help or information to match your particular needs.

References

Bauer, J. L., Kim, Y. J., Zureick, A. H., & Lee, D. K. (2016). *What every science student should know*. Chicago, IL: University of Chicago Press.

Gasiewski, J. A., Eagan, M. K., Garcia, G. A., Hurtado, S., & Chang, M. J. (2012). From gatekeeping to engagement: A multicontextual, mixed method study of student academic engagement in introductory STEM courses. *Research in Higher Education, 53*, 229–261.

Karabenick, S. A. (2003). Seeking help in large college classes: A person-centered approach. *Contemporary Educational Psychology, 28*(1), 37–58.

Karabenick, S. A., & Berger, J. L. (2013). Help seeking as a self-regulated learning strategy. In H. Bembenutty, T. J. Cleary, & A. Kitsantas (Eds.), *Applications of self-regulated learning across diverse disciplines* (pp. 237–262). Charlotte, NC: Information Age.

Karabenick, S. A., & Gonida, E. N. (2018). Academic help seeking as a self-regulated learning strategy. In D. H. Schunk & J. A. Greene (Eds.), *Handbook of self-regulation of learning and performance* (pp. 421–433). New York, NY: Routedge.

Kiewra, K. A. (2005). *Learn how to study and SOAR to success*. Upper Saddle River, NJ: Pearson.

McGuire, S. Y., & McGuire, S. (2015). *Teach students how to learn: Strategies you can incorporate into any course to improve student metacognition, study skills, and motivation*. Sterling, VA: Stylus.

Newman, R. S. (1994). Adaptive help seeking: A strategy of self-regulated learning. In D. H. Schunk & B. Zimmerman (Eds.), *Self-regulation of learning and performance: Issues and educational applications* (pp. 283–301). Hillsdale, NJ: Erlbaum.

Nilson, L. B. (2013). *Creating self-regulated learners: Strategies to strengthen students' self-awareness and learning skills.* Sterling, VA: Stylus.

Szu, E., Nandagopal, K., Shavelson, R., Lopez, E., Penn, J. H., Scharberg, M. A., & Hill, G. (2011). Understanding academic performance in organic chemistry. *Journal of Chemical Education, 88*(5), 1238–1242.

10

Conclusion

While reading this book, you may be thinking to yourself that some of the behaviors and habits that I am suggesting that you adopt as an instructor and mentor of first-generation college students sound somewhat odd or unusual. Or, you may be feeling awkward or uneasy about changing the way you teach.

I would like to reassure you about a number of things:

1. This process is not going to be difficult. You will discover that it is much easier to teach students about study skills, than it is to teach them about some of the more complex or esoteric topics of your discipline.
2. The process can be gradual and gentle. The suggestions that I am making regarding changing your teaching habits do not require radical change. The changes I am suggesting can be made slowly and gradually.
3. While it may seem awkward at first, these new behaviors will soon become automatic. While the teaching and mentoring behaviors which I speak about in this book may initially seem unfamiliar to

Teaching STEM to First Generation College Students, pages 125–126
Copyright © 2019 by Information Age Publishing

you, they will gradually become second nature and will eventually require little to no thought on your part (S. Karabenick, May 22, 2018).

Most importantly however, there is a great payoff at the end. As an instructor, it is not very rewarding to help students when you know that life has always been very smooth and easy for them. But when you see students who overcome large obstacles and challenges and you know that you are an *instrumental* part of their success, this can be most rewarding and heartening. I hope that this book helps bring you the fulfillment and joy that can come from working effectively and successfully with first-generation college students.

About the Author

Gail **Horowitz** received her PhD in science education from Teachers College, Columbia University. She earned her master's degree in organic chemistry at Columbia University where she conducted synthetic organic chemistry research under the direction of Dr. Gilbert Stork.

Dr. Horowitz has had an almost 30 year career in higher education where she has taught a range of chemistry courses, including chemistry for non-majors, organic chemistry for nutrition majors, and organic chemistry for science majors. She has served as an official and unofficial advisor to thousands of students. Dr. Horowitz began her teaching career at a small, private, liberal arts college where she taught for 20 years. She then spent 8 years working at Brooklyn College of the City University of New York (CUNY), a mid-sized, public, urban college that serves a diverse population of immigrant, low-income, minority, and first-generation college students. She is currently teaching at Bard High School Early College in Newark, New Jersey.

Dr. Horowitz conducts primarily action research, practical research looking to improve science teaching and learning. Her current research focuses on pedagogies and interventions that can trigger, foster, and develop the self-regulated learning abilities of organic chemistry students.

CPSIA information can be obtained
at www.ICGtesting.com
Printed in the USA
BVHW042039160719
553634BV00006B/90/P